The ManuFuture Road

Francesco Jovane · Engelbert Westkämper
David Williams

The ManuFuture Road

Towards Competitive and Sustainable High-Adding-Value Manufacturing

 Springer

Prof. Francesco Jovane
Institute of Industrial Technologies
and Automation
Viale Lombardia, 20/A
I-20131 Milano, Italy
Email: f.jovane@itia.cnr.it

Prof. Dr.-Ing. Engelbert Westkämper
Fraunhofer IPA
Nobelstr. 12
70569 Stuttgart, Germany
Email: wke@ipa.fhg.de

Prof. David Williams
Wolfson School of Mechanical and
Manufacturing Engineering
Loughborough University
Leicestershire LE 11 3TU, UK
Email: d.j.williams@lboro.ac.uk

ISBN 978-3-540-77011-4   e-ISBN 978-3-540-77012-1

DOI 10.1007/978-3-540-77012-1

Library of Congress Control Number: 2008936766

Typesetting: Scientific Publishing Services Pvt. Ltd., Chennai, India.
Coverdesign: eStudioCalamar S.L., F. Steinen-Broo, Girona, Spain

Printed in acid-free paper

9 8 7 6 5 4 3 2 1

springer.com

Europe's Manufacturers:
Bringing Together New Ideas with Market Needs

The work of ManuFuture could not come at a better time. Your Strategic Research Agenda, with its ambitious plan to invite European organisations to invest in a set of targeted research, innovation and educational activities, should make a big contribution to our goals. If followed through, it will improve both the competitiveness of, and employment levels in, Europe's manufacturing industries.

It is encouraging to see that ManuFuture has considered all three sides of the knowledge triangle – education, research and innovation. The same thinking went into the Commission proposal for a European Institute of Technology, which will help to close the gaps between our universities, research centres and industry. This is vital if we are going to unleash the full potential of Europe's knowledge economy.

An extract from: "Bringing Together New Ideas with Market Needs".

Porto, 24 July 2006

José Manuel Barroso
President of the European Commission
ManuFuture Industrial Advisory Group

Preface

Manufacturing, covering from products and services, to processes, companies and related business models, is the backbone of European economy. More than 34 million people are employed in more than 2.230.000 enterprises in 23 industrial sectors [1.1]. In the related service areas for manufacturing 60 additional million people are engaged. Nearly 500.000 people are engaged in research, technological development and innovation, related to education, within universites, research institutes and industry. Manufacturing turnover accounts for more than 6.300 BEURO, 55% of the European GDP with an added value of 1630 BEURO [1.2]. Europe is still leading the global trade market.

Key issues, from globalization to climate change, are challenging manufacturing in advanced as well as emerging countries. Hence, manufacturing is getting back to the political agendas and the awareness of stakeholders is rising. In Europe key issues may lead to disruptive changes in the socio-economic system.

The ManuFuture initiative has been promoted to provide: a 2020 Vision, a Strategic Research Agenda (SRA), Roadmaps, awareness of the resources required, basic activities and pilot actions, to help devise and support the European *response* to the key issues challenging manufacturing. These come from the economical, social, environmental and technological (ESET) context changes and call for a move towards sustainable development.

High-Adding-Value (HAV), Knowledge-based (K-based) Competitive Sustainable Manufacturing (CSM), has been proposed by ManuFuture as the European *response*. It would involve all stakeholders, from policy makers, to public authorities and financial institutions, to industry, universities and research institutes and centres. HAV CSM may be seen as the *European Technological and Industrial Revolution for competitiveness and sustainability.*

Pursuing HAV CSM is feasible, as European industry still leads in many domains at global level and the European Education, Research and Technological Development; Innovation (E&RTD&I) System is capable of enabling and supporting a shift to HAV CSM. But human and financial resources should be dedicated. The existing education RTD&I community could be enlarged, as highly educated people are available.

The main problem may be on the side of financial resources and quickness in action. Assuming that the shift to HAV may require 10% of the current investments for continuously upgrading education, the related investment would be in the range of 15

BEURO per year. It would concern from K-based industrial Innovation, though to RTD and education. Following Lisbon strategy, 5 BEURO should be invested by European, national and regional programmes and initiatives. The rest should come from industry.

As time constants, concerning the research-innovation-market value chain, are high, decisions must be taken by the stakeholders and in particular, by politicians, public authorities and industry very soon. It must be acknowledged that Europe is ahead of other global regions and countries. The High-Adding-Value (HAV), Knowledge-based (K-based) Competitive Sustainable Manufacturing (CSM), as proposed by ManuFuture, is being pursued by ongoing European programmes and initiatives. These should be fostered, supported, coordinated and, finally, integrated.

This book addresses the stakeholders and is intended to contribute to their awareness and support their fundamental proactive role and action. The book presents the contribution already given by the ManuFuture initiative, the role this is playing and its further proactive action as well as the European and global evolving economical social environmental technological reference context. Beside ManuFuture activities and results, reference is made to official documents, reported here as closely as possible.

HAV, K-based Competitive Sustainable Manufacturing (CSM) is a revolutionary model of future manufacturing. It refers to studies carried out by the International Academy for Production Engineering (CIRP). CSM covers a wide field from traditional to emerging sectors of industry. It fosters proactive initiatives and concrete fields of actions, to innovate products, processes and enterprises. Pursuing CSM implies transformation of industry, towards HAV, and its supporting knowledge-generating infrastructure: the education, Research and Technological Development and Innovation System, (E&RTD&I).

ManuFuture is an industry-led initiative [1.3] whose mission is to pursue CSM. It aspires to promote investment in innovation that will ensure the future of European manufacturing in a knowledge-based economy. ManuFuture represents a planning and implementation initiative that defines, prioritises and coordinates the necessary scientific technical and economic actions to achieve the objectives set out above. Acting as a new kind of infrastructure, the ManuFuture platform is generating Strategic Intelligence (SI), i.e. the Competitive Sustainable Manufacturing (CSM) Vision, Strategic Research Agendas (SRA) and Roadmaps. It is developing and managing the ManuFuture Framework (FW), where SI is being implemented, to pursue CSM.

This book covers from the anticipated European promoting and supporting activities for sustainable development, to CSM, to the ManuFuture platform activities to generate SI and the framework for SI implementation, to ongoing basic activities and pilot actions pursuing CSM, to future perspectives.

In the first chapter, the move towards Competitive Sustainable Development (CSD) is presented, considering the role of the EU. Then, Competitive Sustainable Manufacturing (CSM), as a fundamental enabler for achieving CSD, is described, with particular reference to its competitiveness and sustainability. The need for a new HAV, K-based manufacturing paradigm and the enabling role of education, research and technological development, innovation – the K-Triangle – is introduced. Then the proactive strategic role of ManuFuture is outlined.

In the second chapter, the role of products and services, processes and companies, in view of pursuing CSM, is analysed. Manufacturing industry situation and perspectives are presented, considering the changes of the global market, the migration of production and consumption, the economic potential of manufacturing as well as the industrial structure, including strengths and weaknesses. Then the European leadership in manufacturing is analysed, referring to customisation, global production and technologies.

In the third chapter, the European ManuFuture initiative is described, covering from the ManuFuture Platform, to Vision 2020 and SRA features: i.e. K-based manufacturing and roadmap for industrial as well as E&RTD&I system transformation, drivers of change, pillars and domains of actions, multi-level action. Further, this chapter reports on the current situation of the E&RTD&I system in Europe and perspective transformation required as emerging from the SRA. Issues concerning investments in RTD are raised here.

In the fourth chapter, following the European way to Competitive Sustainable Development (CSD) manufacturing strategies, in terms of visions, concepts and actions to reach long-term, as well as medium-term, goals and targets, are analysed, referring to products and services, new business models, lean efficient enterprises processes, new ways of working. Innovating manufacturing engineering, from adaptive to reconfigurable manufacturing, to knowledge-based factories as products, to new Taylorism, to networking in manufacturing, is analysed, as well as digital manufacturing engineering. The challenge of advanced industrial engineering, emerging manufacturing technologies and technologies beyond borders is described. The enabler role of manufacturing industries is underlined. This ManuFuture road is a contribution to support industrial strategic planning and work programmes for public trans-sectorial collaborative research. The visions, goals and targets follow the needs of competition and sustainability.

In the fifth chapter, the Roadmaps for manufacturing research, based on the ManuFuture SRA and developed in 2006 and 2007 by the Leadership Consortium (Annex), are reported.

The Roadmaps are driven by industrial and economic requirements and the need for transformation of manufacturing towards CSD. More than 80% of the proposed activities follow visions, strategic objectives and tasks of the ManuFuture pillars and are of common interest for all industrial sectors. The authors summarised them to several trans-sectorial Roadmaps, unified under a comprehensive approach representing the ManuFuture vision towards the European industrial transformation, further on called the ManuFuture work programme.

In the sixth chapter, the ManuFuture road to High-Adding-Value Competitive Sustainable Manufacturing, as emerging from the results achieved and the foreseeable perspectives, is outlined.

ManuFuture, acting as a strategic infrastructure to pursue CSM, has generated SI and the related implementation framework (FW). This encompasses from reference models for action and global cooperation, to EMIRA, to the 25 national ManuFuture platforms and the Knowledge Innovation Community (KIC). Stakeholders, from public authorities and financial institutions, from industry, university, research institutes and centres are cooperating in SI implementation, through basic activities. To speed up and lead the implementation process, pilot initiatives are being explored, developed and launched. A Manufacturing Joint Technology Initiative (JTI) is currently being considered. Its objective is the implementation of manufacturing enabling technologies of the future, on the basis of the ManuFuture SI.

Finally, the Eureka cluster ManuFuture industry, has been conceived. Its definition phase has been launched. It will implement the ManuFuture SI concerning European production systems: products for the world market and processes to retain production in Europe. More than 40 companies, supported by ten research institutes, making up a Knowledge Innovation Community (KIC), will be investing in the cluster. Expected overall value of the projects is 400 MEURO.

The *European Technological and Industrial Revolution for competitiveness and sustainability* can rely on ManuFuture: a contribution to a leading role of Europe.

The ManuFuture road to HAV CSM is the result of four years of activities to find out the way to competition and sustainability. More than 350 actions have been defined and structured in sectorial and trans-sectorial Roadmaps. Nearly 80% of the actions are relevant for all industrial sectors. They are precisely defined and a source of innovations with high economic impact. Proposals for innovating the structure of research and education, the so-called Knowledge Triangle, show the way to the European Manufacturing Innovation and Research Area (EMIRA). Hence, this book may be seen as a 'master' for all experts and people, who are in charge of research and development in enterprises, research organisations and institutes, research foundations, governmental institutions and politics, and those who feel responsible for the development of competitiveness and sustainability in European, national and regional industries.

The authors and co-authors are members of the strategic group which elaborated vision, strategies and Roadmaps. They all spent much of their time for the ambitious goal to formulate the way towards the future of manufacturing in Europe.

The European Commission and some national governments supported the process of roadmapping by a co-ordinated action and by national or regional projects. Without high personal engagement this book could not have reached its high level of detail and concrete actions. The authors, co-authors and the ManuFuture community thank the European Commission and all their colleagues involved for their active support. We thank many experts for their active participation and we hope the book will help to transform manufacturing in Europe to the needs of the future by High-Adding-Value (HAV) and sustainability.

Francesco Jovane	Engelbert Westkämper	David Williams
ITIA-CNR	Fraunhofer IPA	Loughborough
Institute	Fraunhofer Institute	University
of Industrial	for Manufacturing	
Technologies	Engineering and	
and Automation	Automation	

Milano   Stuttgart   Leicestershire

July 2008

Acknowledgement

The development of the European Technology Platform (ETP) ManuFuture was only possible by the support of the Industrial Technologies Directorate of the European Commission's Research DG over several years. The Vision and the Strategic Research Agenda (SRA) have been discussed with stakeholders and experts from research and industries. Several additional co-ordinated actions with initiatives of the ManuFuture Consortium represented a substantial contribution to this book. Especially the co-ordinated action Leadership (NMP2-CT-2006-033416) carried out a considerable amount of work for defining the fields and the road to implementation. Leadership was funded in the Commission's 6th Framework Programme (2002-2006).

During the phases from the Strategic Research Agenda to the Roadmap, many contributions have been integrated: EU-MECHA-PRO, MINAM, which gave special impulses to high-potential fields of research and industrial priorities. Technology Platforms such as Textile, Agriculture, Steel, Aluminum, Rapid Manufacturing, Photonics, Tools, Aerospace, Marine, Chemistry, Forest and others supported the ManuFuture implementation plan and made ManuFuture an umbrella platform for the area of manufacturing in Europe. Not all aspects could be integrated because this book focuses on the core of manufacturing and trans-sectorial activities. Sector specifics can be integrated under this umbrella later. Consultations by ManuFuture with other ETPs in common workshops and bilateral discussions allowed joint initiatives which should be continued under the guidance of the Commission.

Many actions described in this book follow industrial priorities and are proposals for future activities in the government funded research programmes at European, national and regional level. It is the beginning of a trans-European way of networking and co-operation in manufacturing research for competition and sustainability. The high dynamic of innovation in technologies and methodologies make it necessary to rework plans of action on a rolling basis.

Greetings

Professor Heinrich Flegel
President of the ManuFuture High Level Group

Europe has a long and successful tradition in industrial production. In fact industriali-sation started from Europe. Manufacturing is the base of the European economy and welfare. Many sectors of manufacturing have leading positions in the world. But the European manufacturing industries are vulnerable and under attack in the global mar-ket by competitors which operate in regions with lower cost levels.

The European Technology Platform (ETP) ManuFuture is an industrial-driven ini-tiative which follows the Lisbon Agenda for growth and sustainability in the knowl-edge community. A generic model has been defined to change the paradigm from cost-oriented manufacturing to High-Adding-Value (HAV) CSM. Guiding the route and technology in the global market is the strategic objective of more than 23 sectors of industries. Experts from industrial organisations and enterprises, research institutes and universities formulated the route to future development and the Vision of manu-facturing in 2020. ManuFuture is an umbrella platform with many relations to other ETPs and sectorial platforms as well as themes of European research. Research into manufacturing is an investment in the competitiveness of European industries and the employment of more than 34 million people. ManuFuture started as a European initia-tive for research. In the meantime, national and regional platforms followed the strate-gic orientation of ManuFuture in all EU countries. Thus, the forces of research join to a European army of competence and knowledge to accelerate innovation and trans-form the structure of European industries.

The authors of this book did an excellent job in formulating a comprehensive view of research and development for all manufacturing sectors. They elaborated the way and perspectives of future development with strong relations to industrial needs and technological potentials. This book will be the new paradigm for manufacturers and gives industrial management and politicians clear answers for their research invest-ment in the heart of the European economy. Let's now turn vision into business. Let's make it happen!

Carlos Costa
Vice-President of the ManuFuture High Level Group
Presently Vice-President of the European Investment Bank
Former Member of the Board and Executive Director of Caixa Geral dos Depósitos

Manufacturing in Europe is under high pressure in the global market due to a 'pincer effect': on the one side, the need to compete with low-wage competitors that are absorbing very fast the technologies available, and on the other side, the need to keep pace with science-based innovation processes and products that are creating new markets and new business opportunities.

It is therefore crucial to increase Europe's manufacturing productivity and sustainability through new technological developments and, in parallel, to take advantage of the new market opportunities for which demand is growing very fast, creating employment and output opportunities that will compensate for the slow or even declining pace of the more mature markets. If manufacturing is not able to meet this twofold challenge – the key enabling factor for that being science-based innovation – it may lose its central role as Europe's economic base.

The challenge is certainly a company one, but it is very dependent on the contextual environment at regional, national and European levels. Companies are known to be strongly dependent on the nature, efficiency and quality of the regional, national and European innovation systems that encompass three critical sub-systems: education, research and technological transfer and absorption.

All European countries, regions and sectors need a push from innovation focused on High-Added-Value Manufacturing. Investment in research and development in manufacturing has to ensure leadership and competitiveness and is thus mandatory for both industries and communities. The survival of European manufacturing is critical for preserving jobs as well as for giving them high qualification content. This means that the European manufacturing revival is at the heart of the Lisbon strategy inasmuch as it is a pillar of strategic orientation for economic growth, more and better jobs, sustainability and structural change towards the knowledge-based community of this century.

The ManuFuture platform represents the European stakeholders' awareness of the critical role and the urgent need for science, technology and know-how supporting leadership in HAV products, manufacturing processes and business organization.

Over the last four years a Vision and a Strategic Research Agenda (SRA) and Roadmaps have been elaborated, discussed and endorsed by the stakeholders. Industry's high level of engagement throughout Europe and the contributions made by leading European research institutions represent a sound step towards successful development and structural change in Europe.

The way for industry to go now is to make effective the strategic lines which have been defined within ManuFuture. New business models, innovative products end emergent technologies are the expected outcome of the stakeholders' enforcement of the need for a strategy that is built on the Vision and the SRA. Furthermore, ManuFuture does not overlook the need for an efficient use of natural resources, such as energy and raw materials, and encourages building the future of manufacturing on the culture of human work and know-how with their centuries-long European tradition.

I am sure that the road of ManuFuture is of the utmost importance in adding value and making products with European management standards and culture in the world.

The authors and all the contributors to this book should be thanked for their commitment in this endeavour and for being both the brain and the muscle of a Technology Platform that is a cause of general interest. In fact, ManuFuture does not contribute for the benefit of a single group of companies or a specific sector, but rather creates a new positive and winning state of mind and conceptual framework that needs to be spread and shared by all actors that contribute and are decisive for meeting the challenge facing European industry.

This document is a working paper of the services accompanying the green paper on the European Research Area (ERA). It brings together a number of elements supporting the issues raised in the green paper and highlights various facts therein.

Despite the widespread popularity of the ERA concept, there is clearly a need to further deepen the analysis of the performance of the national and European research systems and to assess the implications of the issues and challenges that emerge for ERA. The distribution of, and access to, strategic intelligence among the key policy actors within ERA will be an important tool to satisfy this need, along with a stronger involvement of the academic community in the conceptualisation of ERA.

Contents

Abbreviations

ANN:	Artificial Neural Network
BTO:	Build-to-Order
CAD/CAM:	Computer-aided Design/Manufacturing
CIM:	Computer Integrated Manufacturing
CIP:	Competitiveness and Innovation Framework Programme
CIRP:	The International Academy for Production Engineering
CRM:	Customer Relationship Management
CS:	Competitive/Sustainable
CSD:	Competitive Sustainable Development
CSM:	Competitive Sustainable Manufacturing
DfS:	Design for Sustainability
DG:	Directorate-General
E&RTD&I:	Education, Research and Technological Development and Innovation
E&RTD&I&M:	Education, Research and Technological Development, Innovation and Market
ED:	Economic Development
EMIRA:	European Manufacturing Innovation and Research Area
EMRI	European ManuFuture Research Institute
ERA:	European Research Area
ERP:	Enterprise Resource Planning
ESET:	Economy, Society, Environment and Technology
ET:	Enabling Technologies
ETIR:	European Technological and Industrial Revolution for global competitiveness and sustainability

ETP:	European Technology Platform
FW:	ManuFuture framework
FPs:	EC Framework Programmes
FuTMaN:	The Future of Manufacturing in Europe
GDP:	Gross Domestic Product
GPS:	Global Positioning System
GSM:	Global System for Mobile Communication
HAV:	High-Adding-Value
HCI:	Human-Computer-Interaction
HMM:	Hidden Markov Model
HRST:	Human Resources in Science and Technology
ICT:	Information and Communication Technologies
IEA:	International Energy Agency
I-KTs:	Intelligent Knowledge Triangles
IMD:	International Institute for Management Development
IPP:	Integrated Product Policy
IPR:	Intellectual Property and Rights
ITFSP:	International Task Force for Sustainable Products
KIC:	Knowledge Innovation Community
KIS	Knowledge-intensive Services
KNN:	K-Nearest Neighbour Algorithm
K-Triangle:	Knowledge Triangle
LCA:	Life Cycle Assessment
LKIS :	Less Knowledge-intensive Services
M2M:	Machine-to-Machine
ManVis:	Manufacturing Visions. The specific support action – Integrating Diverse Perspectives into Pan – European Foresight
MATAP:	Action Plan on Manufacturing Technologies
MD:	Molecular Dynamics
MDP:	Markov Decision Process
MEM:	Micro Electro Mechanical System

MES: Manufacturing Execution System

MF.IND :   ManuFuture Industry Cluster

MOEM: Micro-opto-electromechanical System

MPR: Manufacturing Resource Planning

NACFAM: National Council for Advanced Manufacturing

NOE: Network of Excellence

NMP: Nanoscience, Materials and Production Technologies

OEM: Original Equipment Manufacturer

OSA: Open Systems Architecture

PA: Public Authorities

PLC: Power Line Communication

PLC: Product Life Cycle

PMPP: Post Mass Production Paradigm

PPC: Products, Processes and Companies

PS&PR&BM: Products and Services, Processes and Business Models

QTD: Quantity-Time Diagrams

RFP: Research Framework Programme

RFID: Radio Frequency Identification

RTD: Research and Technological Development

RTD&I&M: Research and Technological Development, Innovation and Market

RSE: Researchers

SD: Sustainable Development

SI: Strategic Intelligence

SME: Small and Medium-sized Enterprises

SMP: Sustainable Manufacturing Paradigm

SQ: Sustainable Quality

SRA: Strategic Research Agenda

TEMS: Total Energy Management System

TP: Technological Platform

TPS: Toyota Production System

VR: Virtual Reality

Introduction

F. Jovane

At the turn of the third millennium, rising public awareness of economic, social, environmental and technological problems brings sustainable development and its main enabler, manufacturing, back to political agendas.

Encompassing products, processes, companies and business models, manufacturing is a very relevant wealth generator, job provider and resources user.

Key issues, from globalization to climate change, ageing population, public health, poverty and social exclusion, loss of bio-diversity, increasing waste volumes, soil loss, transport congestion; are challenging manufacturing in advanced as well as emerging countries. Competitive Sustainable Manufacturing (CSM) can be the answer.

Awareness of the political relevance of Competitive Sustainable Manufacturing is growing and political agendas are concerned with fostering and sustaining competitiveness as well as sustainability of manufacturing at world level.

At European Union level, various DGs have been and are dealing with the challenge faced by manufacturing. The Industrial Technologies Directorate of the European Commission's Research DG promoted and supported an initiative concerned with the future of manufacturing that has encompassed

- FuTMan (Future of Manufacturing) [1.4] to MANVIS (Manufacturing Visions – Integrating diverse perspectives into Pan-European Foresight) [1.5]
- the work, in 2003, of an expert group, followed by the ManuFuture 2003 Conference in Milano,
- the launch of the European Technological Platform ManuFuture.

The ManuFuture mission is to pursue High-Adding-Value (HAV), Know-ledge-based (K-based) Competitive Sustainable Manufacturing (CSM), involving the stakeholders, from public authorities and financial institutions, to industry, university, research institutes and centres. To this end, ManuFuture develops, proposes and implements a strategy, based on research and innovation, capable of speeding up the rate of industrial transformation in Europe, securing High-Adding-Value employment and winning a major share of the world's manufacturing output in the future knowledge-driven economy.

Thus, ManuFuture would contribute to the *European Technological and Industrial Revolution for global competitiveness and sustainability.*

ManuFuture, an industry-led initiative, aspires to promote investment in innovation that will ensure the future of European manufacturing in a knowledge-based economy [1.3]. It represents a planning and implementation initiative that defines, prioritises and coordinates the necessary scientific technical and economic actions to achieve the objectives set out above.

A time span of 10 to 15 years can bring about dramatic changes. It is not only industrial labour that is cheaper in some regions outside Europe, but also engineering and management. Today, Europe still has the possibility and means to counteract this

Manu*Future*

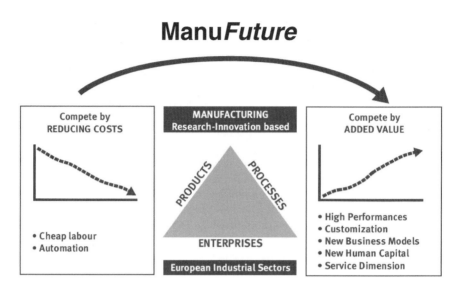

Fig. 0.1. Competition Shift – from Reducing Costs to High-Added-Value: ManuFuture (© ITIA-Series 2004)

situation, but it must do so in a decisive, concentrated manner, based on sound strategic analysis and appropriate investments in Education, Research and Technological Development and Innovation(E&RTD&I).

The ManuFuture Platform must be considered as an infrastructure, whose activities generate:

- Strategic Intelligence (SI): i.e. the Vision for 2020, the Strategic Research Agenda (SRA), the Roadmaps, the new HAV, K-based Competitive Sustainable Manufacturing paradigm, (see figure 0.1.)
- and the ManuFuture framework (FW), encompassing from the reference model for basic activities and pilot actions at European and global level, to the definition of the European Manufacturing Research and Innovation Area (EMIRA); to the knowledge and innovation community, acting within EMIRA, and related reference model for governance; to the 25 national and regional ManuFuture Platforms. It is within this framework that SI is being implemented to pursue CSM.

ManuFuture results, meant to pave the road towards European High-Adding-Value Competitive Sustainable Manufacturing, come at a time when the US, Japan and China are considering manufacturing as a strategic issue.

Hence, implementation of the results, achieved so far by ManuFuture, is of paramount importance as well as increasing political awareness concerning European High-Adding-Value industry. The main issues, dealt with in the following chapters, may be summarised as follows:

- moving towards CSM and proactive strategic role of ManuFuture,
- the role of products and services, processes, enterprises and business models, in view of CSM; situation and prospects of European industry,

- the ManuFuture Initiative and the situation and prospects of E&RTD&I,
- the analysis of manufacturing strategies, goals and targets, for competitiveness and sustainability,
- the Roadmaps for manufacturing research, based on ManuFuture SRA, driven by industry requirements and needs to pursue CSM,
- the generation of SI and its implementation within ManuFuture framework, to achieve CSM, through basic activities and pilot actions.

1

Leadership of European Manufacturing Industry

F. Jovane

Competitive and Sustainable Development (CSD) is emerging as the necessary global strategic vision to be implemented to meet the economic, social, environmental, technological challenges we all face at various levels from global to local level. The European Union started fostering and supporting such process long ago. It has been developing a "rolling" strategic sustainability policy that crosses any initiative at European level. Member states are taking a similar approach.

In this chapter, the move towards Competitive Sustainable Development is presented, considering the role played by the EU. Then, competitive sustainable manufacturing, as a fundamental enabler for achieving CSD, is described, with particular reference to its competitiveness and sustainability. The need of a new HAV, K-based, Competitive Sustainable Manufacturing (CSM) paradigm and the enabling role of education, research and technological development, innovation – the K-Triangle – are introduced. Then the proactive strategic role of ManuFuture is outlined.

1.1 The Emerging Competitive Sustainable Development Paradigm: Towards a World Initiative

The first industrial revolution was the starter and enabler of a long lasting economic growth, then development, paradigm based on innovation. But it was accompanied by social disasters and migration of manufacturing to other regions of Europe. Such a paradigm was adopted, through time, by different countries leading to an economic growth and development, unprecedented in history. A new European way for balancing work and capital was achieved through the establishment of workers unions. Social regulations and political changes followed over the last centuries.

The economic development paradigm has been affecting – and also is affected by – the economy, society, environment and technology (ESET) context. Only in the sixties, the "Club of Rome" by promoting the study: "The limits to growth", opened the way to the sustainable development paradigm [1.7]. In the eighties, sustainable development (SD) – concerned with economic, social and environmental development – was considered the goal of a desired new industrial revolution, involving advanced and new emerging countries. This goal has not been achieved, mainly due to the strong focus on competitiveness and profitability that has been characterising the emerging globalisation. Globalisation has activated a new industrial revolution, enabled and supported by new technologies, that is leading to a new world distribution of production and markets. It is affecting all countries and economic regions. A fierce competition has and is taking place among advanced and new emerging countries.

Hence, economic development (ED) based on competitiveness is still the dominating paradigm. Most of our manufacturing policies and RTD efforts address ED.

Economic, social and environmental problems are developing very fast and must be solved. Figure 1.1 shows a great challenge: the world's unbalance concerning energy consumptions and materials use, mainly due to migration of production. It is mandatory for advanced as well as emerging countries to define, move towards and implement a global competitive sustainable development (CSD) paradigm, see figure 1.2,

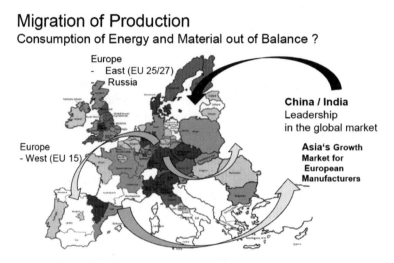

Fig. 1.1. Migration of Production: Consumption of Energy and Materials out of Balance? [1.8] ©IFF/IPA

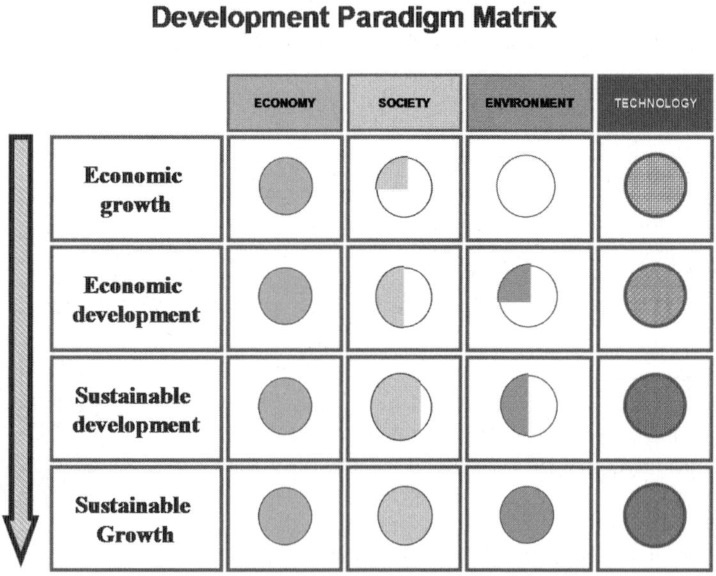

Development Paradigm Matrix

	ECONOMY	SOCIETY	ENVIRONMENT	TECHNOLOGY
Economic growth	◔	◔	○	◔
Economic development	●	◑	◕	●
Sustainable development	●	●	◑	●
Sustainable Growth	●	●	●	●

Fig. 1.2. Towards Competitive Sustainable Development and Further [1.9]

complying with advanced as well new emerging countries conflicting expectations and interests and with economy, society, environment, technology (ESET) context requirements. The term sustainable here includes economic, social, environmental and technological sustainability.

Competitive sustainable development is a long-term vision. Its main enabler is Competitive Sustainable Manufacturing (CSM). This depends on and affects:

- manufacturing: i.e. products and services, processes, business models,
- related policies concerning: education, research and technological development, industrial innovation.

Appropriate paradigms and general reference models have been and are being developed.

The first paradigm to be proposed [1.10] was the "dynamic model of the sustainable production". According to it, the product-processes life cycle matrix evolution is driven by the economy, society (including environment) and technology context, while RTD&I activities would act as enablers of such evolution. It is this approach that has led to ManuFuture [1.11, 1.12]

A general reference model, see figure 1.3, has been developed and is being proposed [1.13]. It has been developed, starting from Schumpeter theory, assuming that economic development depends on clusters of innovations and referring to Kondratiev long waves or business cycles.

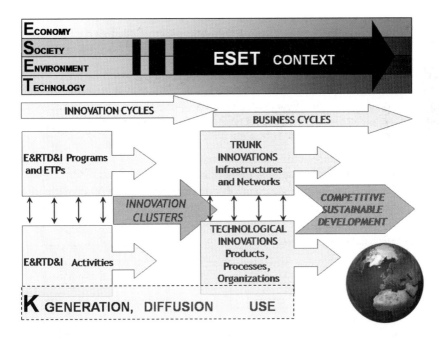

Fig. 1.3. Reference Model: from Innovation Cycles to Business Cycles and Sustainable Development [1.13]

The reference model connects:

- innovation cycles, based on support initiatives (E&RTD&I Programmes and Initiatives, at various levels), acting as infrastructures, and Enabling Technologies (ETs) generation and diffusion,
- to business cycles, based on trunk innovations (infrastructures) and technological innovations; these being based on ETs adoption and use phase,
- to manufacturing paradigms,
- and finally to the development paradigm: in our case the Competitive Sustainable Development paradigm.

All this relies on the K-generation, diffusion, adoption and use process. It integrates innovation and business cycles, covering from science to market, and complies with the economy, society, environment and technology (ESET) context.

Four interlinked levels of intervention should be considered:

- World
- National
- Regional
- Local

To pursue CSD a new K-based Competitive Sustainable Manufacturing paradigm must be developed and implemented, in each country and world region. Co-operation among countries and regions, particularly on sustainability, must take place.

1.1.1 From Competitive Sustainable Development to Competitive Sustainable Manufacturing

Action for devising and implementing the CSM paradigm must take place, involving all the stakeholders, from public authorities and financial institutions, to industry, university, research institutes and centres. This requires a co-shared Strategic Intelligence (SI) – see chapter 1.4 – and, by each stakeholder concerned, a proactive behaviour that takes into account the following dimensions:

- the economic, social, environmental and technological context,
- the geographic scale, from global to local and vice versa,
- the stakeholders to be involved, from PAs to institutions, to individual entities,
- the time horizon, from medium-long to short term,
- the processes to be activated and their "governance",
- the human as well as financial resources involved.

A highly reactive behaviour should complement the proactive behaviour, as everything will rapidly evolve.

1.1.2 Key Challenges

Key challenges, as reported hereafter, are already acting.

- *Globalisation* could jeopardize CSD. But its consequences – from mutual vulnerability to the spread of a transnational consumer class, China syndrome, resource-light wealth, cosmopolitan politics – may work at the end in favour of CSD [1.14].

- *Climate change* is affecting physical and biological systems, shows ethical issues with profound demographic global equity and security consequences.
- *Ageing population.* To cope with the demographic change and with the ageing population in many western European countries, new forms of work organisation are to be developed to enable the full integration of the elderly workforce, including the integration of low qualified men and women.
- *Public health.* Severe threats to public health are posed by new antibiotic-resistant strains of some diseases and, potentially, the longer-term effects of the many hazardous chemicals currently in everyday use; threats to food safety are of increasing concern.
- *Poverty and social exclusion* have enormous direct effects on individuals such as ill health, suicide, and persistent unemployment. The burden of poverty is borne disproportionately by single mothers and older women living alone. Poverty often remains within families for generations.
- *Loss of bio-diversity.* Animal stocks in several countries are near collapse.
- *Waste volumes* have persistently grown faster than GDP.
- *Soil loss and declining fertility* are eroding the viability of agricultural land.
- *Transport congestion* has been rising rapidly and is approaching gridlock. This mainly affects urban areas, which are also challenged by problems such as inner-city decay, sprawling suburbs, and concentrations of acute poverty and social exclusion.

1.2 The European Union Action

Sustainable development concerns "a society which delivers a better quality of life for us, for our children, and for our grandchildren". It implies the integration of the economic, social and environmental dimensions of our society, and "requires that economic growth supports social progress and respects the environment, that social policy underpins economic performance, and that environment policy is cost-effective" [1.15]. Several actions towards sustainability are being taken at various levels on a global scale both by advanced and emerging countries.

The Japanese government's Third S&T Basic Plan 2006-2010 underscores the necessity to achieve sustainable economic growth based on environmental protection and the constant creation of innovation [1.16].

The US Department of Commerce aims to develop a strategy, designed to foster competitiveness in manufacturing and stronger economic and continuous growth at home and abroad, by launching the "Manufacturing Initiative" [1.17].

China has set out her development policy guidelines in the 2006 Five Year Plan that puts emphasis on the development of a harmonious society in which more consideration is given to the social implications that are associated with rapid economic development [1.18].

The European Union started fostering sustainable development in the 90s. It became a fundamental objective, enshrined in its treaty [1.19].

The European Council in Göteborg (2001) adopted the first EU Sustainable Development Strategy (SDS) [1.15]. It was complemented, by an external dimension, in 2002 by the European Council in Barcelona in view of the World Summit on

Sustainable Development in Johannesburg (2002). The external dimension is represented by the key challenges reported before.

Since these negative trends bring about a sense of urgency, short-term action is required, whilst maintaining a longer-term perspective. The main challenge is to gradually change our current unsustainable consumption and production patterns and the non-integrated approach to policy-making.

Against this background the European Council [1.20] has adopted an ambitious and comprehensive renewed SDS for an enlarged EU, building on the one adopted in 2001.

This document sets out a single, coherent strategy on how the EU will more effectively live up to its long-standing commitment to meet the challenges of sustainable development, concerning economy, society, environment, technology.

It reaffirms the need for global solidarity and recognises the importance of strengthening our work with partners outside the EU, including those rapidly developing countries that will have a significant impact on global sustainable development.

The overall aim of the renewed EU SDS is to identify and develop actions to enable the EU to achieve continuous improvement of quality of life; both for current and for future generations, through the creation of sustainable communities able to manage and use resources efficiently and to tap the ecological and social innovation potential of the economy, ensuring prosperity, environmental protection and social cohesion [1.20].

1.2.1 European Union and Competitiveness

Several EU initiatives have referred to and are referring to competitiveness. Following the Lisbon Strategy promoting growth and employment, the Commission has recently proposed a relevant programme for the competitiveness of the EU and its businesses: the Competitiveness and Innovation framework Programme (CIP). This will bring together into a common framework specific community support programmes and relevant parts of other community programmes in fields critical to boosting European productivity, innovation capacity and sustainable growth, whilst simultaneously addressing complementary environmental concerns [1.21].

CIP places the emphasis on open markets. Fundamental to the programme are the rejection of protectionism in Europe, the opening up of the principal markets outside Europe and the bringing together of the EU's internal and external policies. The programme aims at strengthening the EU's external competitiveness and meeting global challenges. To achieve this, the action plan identifies the necessary priorities and methods, comprising an internal and an external dimension.

As for the internal dimension, European businesses must benefit both from the EU's competitiveness based on sound internal policies and opening up of foreign markets. European citizens should feel the advantages.

As for the external dimension the EU maintains its commitment to multilateralism. This offers the means to eliminate trade barriers in a stable and sustainable manner [1.21].

1.2.2 European Union and Sustainability

In 2006 the European Council adopted a renewed sustainable development strategy (SDS) that sets out a coherent plan on how the EU will more effectively respond to the principles and the overarching objective of sustainable development enshrined in the Treaty. The plan consists of seven key challenges and connected cross cutting policies [1.22].

Key challenges encompass from climate change and clean energy to sustainable transport, sustainable consumption and production, conservation and management of natural resources, public health, social inclusion, demography and migration and global poverty.

Cross cutting policies concern from education and training to research, technology and development, financing and economic instruments. For Europe to move along a sustainable development path and maintain current levels of prosperity and welfare [1.23] the above key challenges and cross-cutting policies must be tackled. To this end it must be recognised that SDS goals can only be met in close partnership with the member states. To measure progress towards sustainable development, in February 2005 the European Commission adopted a set of sustainable development indicators [1.23] that refer to figure 1.2. As an integral part of the European strategy, they take into account economic development, poverty and social exclusion, ageing society, public health, climate change and energy, production and consumption patterns, management of natural resources, transport, good governance and global partnership.

1.2.3 A Fundamental Enabler of CSD: Competitive Sustainable Manufacturing (CSM)

Manufacturing – encompassing from products and services, to processes sustaining their life cycles, to companies and business models – is a fundamental enabler of CSD.

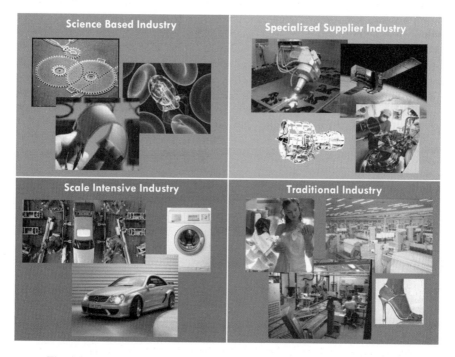

Fig. 1.4. Manufacturing Industry: the backbone of European Economy [1.24]

It generates wealth, sustains jobs directly and through related services, is strictly related to economy, society, environment and technology: it is a fundamental player for sustainability (figure 1.4).

European Manufacturing and related E&RTD

European industry is made up of around 430.000 European manufacturing enterprises with 10 and more employees providing 28 million people with jobs. The turnover of the 23 sectors – making up manufacturing – is 6300 BEURO.

The value added by manufacturing amounting to 1630 BEURO [1.25]. Some 70% of this total was derived from six main areas – automotive engineering, electrical and optical equipment, foodstuffs, chemicals, basic and fabricated metal products, and mechanical engineering [1.26]. EU manufacturing industry provides around 20% of EU output [1.27] and employs some 34 million people [1.25] (figure 1.5).

Over 80% of EU private sector RTD expenditure is spent in manufacturing. This generates new and innovative products that provide some 75% of EU exports [1.27]. SMEs make up the largest part of manufacturing: over 99% of companies are SMEs. They provide 58% of manufacturing employment. Manufacturing accounts for around of one fifth (18.3%) of total value added in EU-25 [1.27].

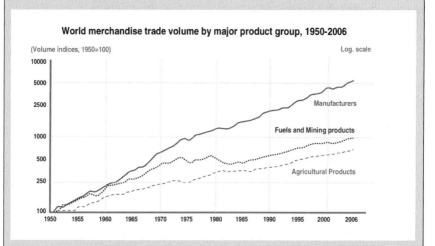

World merchandise trade volume by major product group, 1950-2006

Fig. 1.5. World Merchandise Trade Volume by Major Product Group (1950-2006) [1.28]

The highest share of EU exports (34%) is accounted for by sectors which are characterised by low labour skills (food, drink and tobacco, textiles, clothing, leather and footwear, rubber and plastics, non-metallic mineral products, basic metals, motor vehicles, furniture, miscellaneous manufacturing; recycling), although exports of high labour skills products also account for a significant (27%) part of EU sales abroad [1.25] (figure 1.6).

In terms of technological content of the products traded, the distribution of EU-25 exports is balanced, with a higher revealed comparative advantage in medium-high technology products.

Furthermore, a relative improvement in the EU performance in high technology products has taken place over time [1.29]. Manufacturing took the largest share with an EU-25 average of 81% of total business RTD expenditure [1.30].

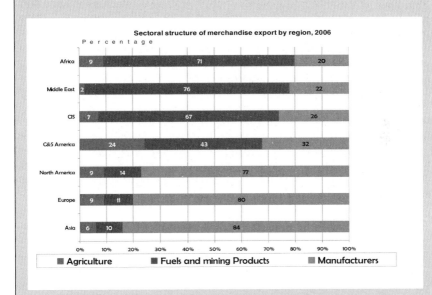

Fig. 1.6. Sectorial Structure of Merchandise Exports (2006) [1.24]

In 2003, in EU-25, 573 000 RTD researchers, measured in FTE, were employed in the business enterprise sector. The largest share of these business RTD researchers (413 000) were working in the manufacturing sector.

> In 2004, the government budget appropriations on RTD, for objectives concerning industrial production and technology, account for 11.4% in EU 15, while in Japan and United States they only account for 7.1% and 0.4% respectively [1.31]. EU countries still produce a greater share of science and technology graduates than Japan or the United States, despite the smaller share of researchers in the workforce: 27% of EU university graduates obtain a science or engineering degree compared to 24% in Japan and just 16% in the United States.
>
> The EU also produces more PhD graduates than the United States, which for its part offers more post-doctoral positions (46.716 in 2003), more than half of which go to foreign PhD graduates. In 2003, over 14 million people in the EU were following tertiary education courses, more than 350,000 were PhD students. One student in four, in 2003, was following a course either in science, mathematics and computing or in engineering, manufacturing and construction [1.31].

Manufacturing concerns from macro (macro-economics) to meso (production and consumption paradigms) to local level (products/services, processes, business models).

Manufacturing involves a large variety of stakeholders, from public authorities (PAs) and financial institutions, to industry, to the education & research and technological development & innovation (E&RTD&I) system. This makes up the Knowledge Triangle (K-Triangle)(figure 1.7).

Fig. 1.7. The Knowledge Triangle [1.9]

It refers to the interaction between research, education and innovation, which are key drivers of a knowledge-based society. In the 90s, following the sustainable development (SD) vision, the sustainable manufacturing paradigm was introduced [1.18]. Due to the strong focus on competitiveness and profitability, that was, at that time, characterising the emerging globalisation, sustainability was somehow set apart. As stressed by the conference "Sustainable Neighbourhood – from Leipzig to Lisbon 2007", sustainability is an engine for European competitiveness within the Lisbon Agenda [1.32]. Manufacturing must change from short-time profit orientation towards long-term sustainability of enterprises, innovative products and processes and reduction of energy and material consumption. Sustainability is not only a factor of cost but also a source of value and innovation. Sustainability should be part of the European way of manufacturing and influence enterprise management, product design and processes [1.33].

Taking the emerging global technological and industrial revolution as an opportunity rather than a threat, industry should develop and implement Competitive Sustainable Manufacturing (CSM). This would enable and sustain global Competitive Sustainable Development (CSD). European industry must make such a move while facing a great challenge in mature as well as advanced sectors. Hence, it must become sustainable while maintaining its competitiveness [1.34, 1.35]. The necessary competitiveness and sustainability should be achieved by developing and implementing new HAV, K-based products and services, processes, companies and business models.

This is a transformation of industry that must be supported by the European E&RTD&I (the K-triangle) system, that – in turn – should become more and more robust and effective, competitive and sustainable, while getting global.

1.2.4 Competitiveness and CSM

Competitiveness in broad terms is a complex concept and may be dealt with at macro, meso, micro level.

Competitiveness at macro level

Country competitiveness was introduced by Porter (1990) [1.36]. In his fundamental work on the competitive advantage of nations, Porter identifies competitiveness of nations as based on the productivity with which it produces goods and services. Competitiveness is of paramount importance for implementing CSM.

This emerges clearly from the various approaches used to assess it. The world competitiveness assessment of the Institute for Management Development (IMD) is based on the assumption that an economy's competitiveness cannot be reduced only to GDP and productivity because enterprises must also cope with political, social and cultural dimensions [1.37].

IMD ranking makes use of four basic constellations of criteria: i.e. economic performance, government efficiency, business efficiency, infrastructures. 55 economies have being ranked. Trend and competitiveness perspectives are being produced.

The World Economic Forum [1.38] performs assessment of countries competitiveness related to their move towards competitive sustainable development, where sustainability is concerned with ESET. The global competitiveness index and consequent ranking refers to the development paradigm matrix, figure 1.8, and the following pillars [1.39]. Institutions, infrastructure, macro economy, health, primary education, higher education, training, market efficiency, technological readiness, business sophistication, innovation Figure 1.8 shows the ranking of countries.

Competitiveness at meso level

At industrial sector level, competitiveness may be described as the capacity to grow, to innovate and produce more and better (higher quality) goods and services, and, hence, to gain market shares in international and domestic markets [1.40].

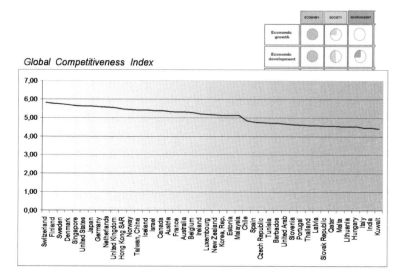

Fig. 1.8. Country Ranking by Global Competitiveness Index [1.39]

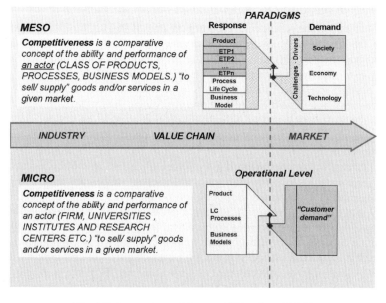

Fig. 1.9. Innovation at Meso and Micro Level [1.9]

Further, competitiveness can be seen as comparative concept of the ability and performance of a supply paradigm to respond to a demand paradigm. The evolution of production paradigm, see figure 1.10 has complied with the above. The performance indicator, proposed by Hon, gives indications about competitiveness of manufacturing systems, as shown in figure 1.11.

Paradigm	Craft Production	Mass Production	Flexible Production	Mass Customisation and Personalisation	Sustainable Production
Paradigm started	~1850	1913	~1980	2000	2020?
Society Needs	Customised products	Low cost products	Variety of Products	Customized Products	Clean Products
Market	Very small volume per product	Demand > Supply Steady demand	Supply > Demand Smaller volume per product	Globalization Fluctuating demand	Environment
Business Model	Pull *sell-design-make-assemble*	Push *design-make-assemble-sell*	Push-Pull *design-make-sell-assemble*	Pull *design-sell-make-assemble*	Pull *Design for environment-sell-make-assemble*
Technology Enabler	Electricity	Interchangeable parts	Computers	Information Technology	Nano/Bio/ Material Technology
Process Enabler	Machine Tools	Moving Assembly Line & DML	FMS Robots	RMS	Increasing Manufacturing

Fig. 1.10. Evolution of Demand and Related Response Paradigms [1.41]

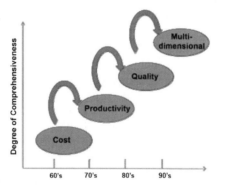

Fig. 1.11. Evolution of Performance Measures for Manufacturing Systems [1.42]

Competitiveness at micro level

Competitiveness at micro level, see figure 1.9, may be defined as a *comparative concept*: i.e. the ability and performance of an actor: from firms to universities, institutes and research centres, etc.; to respond to a "customer demand" better than any other one. Competitiveness at meso and micro level will evolve according to sustainable development requirements. Such evolution may need, at meso and local level, E&RTD&I activities concerning both competitiveness and sustainability.

1.2.5 Sustainability and CSM

Sustainable development may be dealt with at macro, meso, micro level.

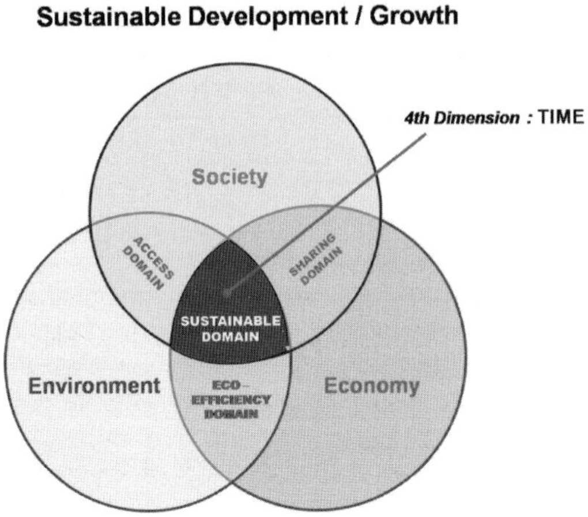

Sustainable Development / Growth

Fig. 1.12. Fundamentals of Sustainable Development [1.8]

At macro level

Sustainable development is a complex concept, concerning three domains – economy, society, environment, and their interactions, as shown in figure 1.12. Several definitions have been proposed.

The world Commission on Environment and Development declaration reads: "Sustainable development is a process of change in which the exploitation of resources, the direction of investments, the orientation of technological development, and institutional change are made consistent with the future as well as present needs" [1.43]. Elkington [1.44], considering enterprises, introduced the 'triple bottom line' concept. This refers to meeting economic goals (profits) and, simultaneously, meet environmental and social goals (or 'bottom lines') in carrying out business. Sustainable value creation is the key contribution of corporations to sustainability; that is, to create long-term value on an economically, socially and environmentally sustainable basis. According to Seliger [1.45] sustainability is directed to enhancing human living standards while improving the availability of natural resources and ecosystems for future generations. More than half of global value creation of today is achieved by less than one tenth of the global population. A sustainable political, economic and social stability can only be achieved if mankind is able to create worldwide – and not only in the first world – jobs and living conditions of human dignity.

Engineering challenges derived from this requirement are a design of products and processes with improved usefulness and less environmental harm. Technology, interpreted as science, systematically exploited for purposes, offers huge potentials. Technology enables processes transforming natural resources into products to meet human needs. The interaction between research and education imposes dynamics on how creative solutions are developed for relevant tasks. Human activities have a high impact on the natural environment, the ecosystem. Reducing such impact, while fulfilling present needs, is a challenge. Living and physical environment, natural resources, processes and balances, must be secured. It is also necessary to reduce global warming, stop loss of biodiversity, control and limit emission of pollutants, take action against many more rising challenges [1.46]. Economic growth is fundamental for sustainable development, as it may enable positive actions and investments within the aforementioned domains, but it must take place with reduced environmental impact. This and economic growth must be decoupled. Improved eco-efficiency can contribute. Further, economic policy and market mechanisms must support sustainable development. Costs for any activity, including long-term environmental and social costs, must be reflected directly in market prices. Markets, where environmental goods and services are traded at real cost, should be established [1.47].

Because of its social dimension, sustainable development, aims at enlarging the vision from the present needs towards the future generations' expectations. Sustainable development can be achieved, as people feel that they can have a fair share of wealth, safety and influence. It is not individual gain, but equitable growth for everybody. Supporting society, involving it in decision processes at various levels make the social dimension of sustainable development. Fighting against poverty through employment, support to sustainable livelihoods, antidiscrimination and social security are a relevant part of the social dimension of sustainable development. Promoting sustainable consumption and production by addressing social and economic development within the carrying capacity of ecosystems and decoupling economic growth from environmental degradation is thus vital. Sustainability in manufacturing requires that key challenges be met by innovative K-based solutions (figure 1.13).

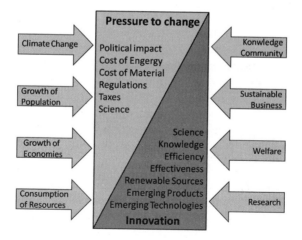

Fig. 1.13. Sustainability in the Focus of Globalisation [1.8]

At global and national level sustainability assessment are carried out using different approaches, as reported below. The European Commission [1.23] is using a set of indicators for monitoring the implementation of its sustainable development strategy. This activity is primarily aimed at providing a first progress report on the current state of play in the implementation of the strategy.

The trends derived from the analysis of indicators are assessed against policy objectives to inform the general public and decision makers about achievements, trade-offs and failures in attaining the commonly agreed objectives of Sustainable Development. Activities pursuing CSM may benefit from the above.

At meso level

For Competitive Sustainable Manufacturing (CSM) to be a relevant enabler of CSD, appropriate paradigms must be devised. They would concern products and services, processes and business models.

Sustainable manufacturing must respond to

- economic challenges, by producing wealth and new services ensuring development and competitiveness through the time,
- social challenges, by promoting social development and improved quality of life through renewed quality of wealth and jobs,
- environmental challenges, by promoting minimal use of natural resources (in particular non renewable) and managing them at the best while reducing environmental impact.

Sustainability of products services, processes, companies and business models may be described as follows [1.48](figure 1.14):

- safe and ecologically sound throughout their life cycle,
- designed to be durable, repairable, readily recycled, compostable, or easily biodegradable,
- produced and packaged using minimal amounts of most environmentally benign materials and energy.

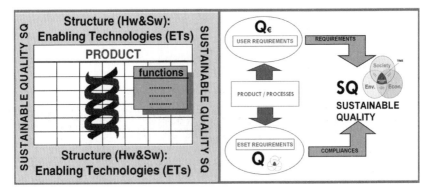

Fig. 1.14. Product Matrix and Sustainable Quality [1.49, 1.50]

Processes (figure 1.15) must be designed and operated so that

- wastes and ecologically incompatible by-products are continuously reduced, eliminated or recycled on-site,
- chemical substances or physical agents and conditions that present hazards to human health or the environment are continuously eliminated,
- energy and materials are conserved, and the forms of energy and materials used are most appropriate for the desired ends,
- work spaces are designed to continuously minimize or eliminate chemical, ergonomic and physical hazards.

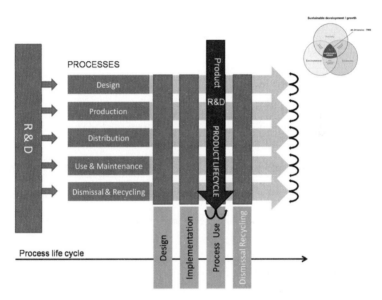

Fig. 1.15. Product - Processes Life Cycle Matrix [1.10]

Workers

- are valued and their work is organized to conserve and enhance their efficiency and creativity,
- their security and well-being is a priority,
- they are encouraged and helped to continuously develop their talents and capacities,
- their input to and participation in the decision making process is openly accepted.

Communities related to any stage of the product life cycle – from production of raw materials through manufacture, use and disposal of the final product – ought to be respected and enhanced economically, socially, culturally and physically.

At micro level

A prerequisite for all sustainability is a strategy that accepts the company's responsibility and its vital role in every society it operates in, and also in the global environment.

For the business enterprise (figure 1.16) sustainable development means adopting business strategies and activities meeting the needs of the enterprise and its stakeholders today, while protecting, sustaining and enhancing the human and natural resources that will be needed in the future. In order to implement sustainable development in manufacturing business enterprises a sustainable manufacturing strategy must be pursued. Sustainable manufacturing (figure 1.17) will picture the idea of sustainable enterprise and the triple bottom line [1.51].

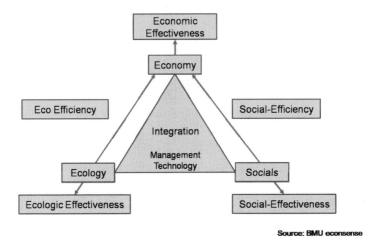

Source: BMU econsense

Fig. 1.16. Challenges of Sustainable Enterprises [1.39]

Sustainability in manufacturing seems to be a cost driver and antipodal to competitiveness. But the reduction of the consumption of resources like energy or material or the dematerialization of product functionality reduces costs. There are already many best practices showing the economic benefits and the potential of adding value. The reduction of resources consumption is a strong contribution to economic success. Some companies have successfully included sustainability into their strategic orientation and are now seeing the results in terms of profitability and market position. [1.52].

Another dimension is the conflict between orientation towards innovation by high technologies and positioning the enterprises in the high-end against the conventional and, low technologies or, on the other hand, low technology products for emerging markets.

The CSD paradigm changes industrial strategies by integration of sustainable manufacturing aims into the overall production system and by the implementation of innovative technologies for overcoming the existing limits in all areas and processes over the life of products, factories and their resources. [1.52]. The requirement is to implement sustainability into the enterprise's strategic plan and to maximize the economic, ecological and social benefits of each technological product during its life cycle. It is even necessary to investigate RTD and technologies according to the above mentioned objectives and to activate their technological potential. [1.52]. Leadership is essential to the successful integration into manufacturing. Partnership is the key. Each stakeholder brings a unique perspective that, when joined with other perspectives, can lead to successful problem solving. [1.52].

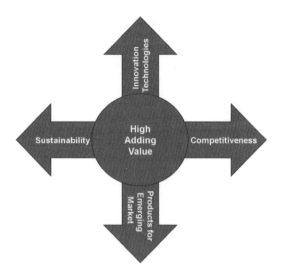

Fig. 1.17. Sustainable Manufacturing [1.4]

1.2.6 Sustainable Quality in Competitive Sustainable Manufacturing

Considering that products as well as processes and enterprises must respond to market needs and comply with sustainability requirements, the term Sustainable Quality (SQ) [1.49] should be used. The product's generation, use and dismissal throughout the product life cycle, is shown in figure 1.18.

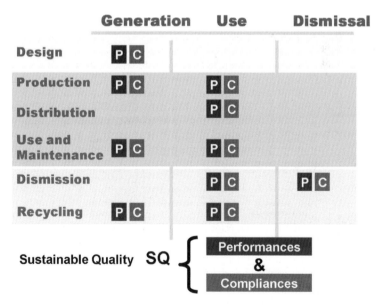

Fig. 1.18. Sustainable Quality (SQ): Generation-Use-Dismissal throughout Product Life Cycle [1.49]

At product-process level

At product-process level the sustainable quality demand curve representing the evolution of the ESET context is shown in figure 1.19 (left part).

Products SQ should evolve through time to respond to rising economic, social, environmental, technological (ESET) context demand curve, as shown in figure 1.19, left. Such evolution, as shown in the right part of figure 1.19, follows two main paths: stepwise improvement and incremental innovation or E&RTD-based innovation. A similar approach may be followed for processes.

The perspective sustainable quality demand or potential curve (figure 1.19) helps to identify either enabling technologies required or the potential use of new emerging technologies. All this is related to E&RTD&I activities.

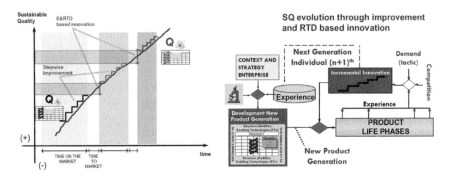

Fig. 1.19. Product SQ Evolution through Improvement and E&RTD-based Innovation to match the SQ ESET Context Demand Curve [1.49]

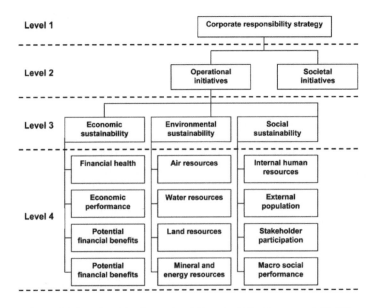

Fig. 1.20. Levels of the Operational Sustainability Framework [1.53]

At company level

At local level sustainability concerns products, services and business models. They should be developed by the stakeholders concerned, complying with the requirements of HAV, K-based Competitive Sustainable Manufacturing.

Development of sustainable products and services, processes and business models at company level may rely on assessment methodologies and tools, as reported hereafter. Sustainability paradigms at meso level will be evolving quite rapidly.

They should be adopted by industry and other stakeholders concerned, shortening the time to market. The incorporation of sustainable development in business practices is of great operational relevance. Referring to figure 1.20, several levels should be concerned: i.e. the strategic level, process or methodological level and operational level. Drivers for such incorporation are showed in figure 1.21.

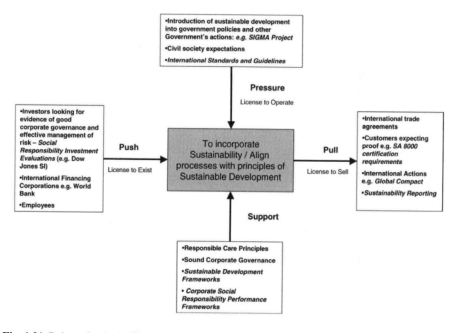

Fig. 1.21. Drivers for the Incorporation of Sustainable Development in Business Practices [1.53]

1.3 A Quest for a New Manufacturing Paradigm: The Role of ManuFuture

The previous analysis shows that European industry must move towards the new paradigm: HAV, K-based, CSM, while facing a great challenge in mature as well as advanced sectors, respectively, by new emerging and advanced countries.

The new paradigm should be developed and implemented:

- at various level, from macro-economics to local level,
- within different economic, social, environmental, technological conditions,

- involving, at various levels, all the stakeholders: from PAs, to financial institutions, to industry, academia and research institutes.

Competitiveness and sustainability, as discussed before, can only be achieved by developing HAV, K-based products and services, processes, companies and business models: i.e. transforming industry at fast rate.

This, in turn, needs a strong and continuous involvement of the European E&RTD&I system, that should become competitive and sustainable, at global level: see figure 1.22.

Fig. 1.22. Factors for a Competitive and Sustainable E&RTD&I System [1.6]

Then both industry and E&RTD&I system should undergo specific transformation processes that would take place respectively within:

- internal and global market,
- EMIRA, the E&RTD&I internal market, getting global.

The two transformation processes and domains concerned must be interrelated. Their effectiveness in enabling and sustaining required transformations, would depend on having a vision, a reference model, proactive enablers ensuring implementation and governance while moving towards the new HAV K-based CSM paradigm, and even further.

As stated before, the above may be seen as the European Technological and Industrial Revolution for global competitiveness and sustainability. Being K-based, as fostered by the Lisbon strategy, it greatly relies on the E&RTD&I system.

Such revolution is fundamental for European manufacturing industry, considering its relevance, see box "Manufacturing" in Chapter 1.3.

Its governance and effectiveness require from a robust reference model, to strategic intelligence (Vision, SRAs and Roadmaps) to the involvement of all stakeholders, to investment in human and financial resources, to a rolling process covering from monitoring to generating new SI for action, to a strong interaction with the globalising E&RTD&I System.

ManuFuture provides answers to such crucial needs. They contribute to the development or revision by PAs as follows:

- policy in terms of: RTD goals, instruments, and procedures of public and private programmes,
- policy measures concerning framework conditions for innovation (science-society relations, industrial relations, human resources mobility, IPR, etc.).

They are of strategic interest for all stakeholders, from companies to universities, research institutes and centres as well as financial institutions.

References

[1.1] Nace rev. 0.1 Classification – ATECO classification (2007)
[1.2] Eurostat Structural Business Statistics. Annual National Accounts (2005)
[1.3] http://www.Manufuture.org
[1.4] EC – DG Research; The Future of Manufacturing in Europe 2015-2020: The Challenge for Sustainable Development (FuTMan) (2002)
[1.5] MANVIS (Manufacturing Visions – Integrating diverse perspectives into Pan-European Foresight, http://www.manufacturing-visions.org/ManVis_Report_2_Final.pdf
[1.6] Report of the High level Group: ManuFuture Vision for 2020 – Assuring the Future of Manufacturing in Europe (November 2004)
[1.7] Meadows, D., Meadows, D.L., Randers, J., Behrens, W.W.: The Limit to Growth. Universe Book (1972)
[1.8] Westkämper, E.: ManuFuture Converence Porto, Pt (2007)
[1.9] Jovane, F.: How to Maintain Competitiveness on the way towards Sustainable Manufacturing. Tampere Manufacturing (summit, June 6-7, 2007)
[1.10] Jovane, F.: The impact of new manufacturing paradigms on society and working conditions. In: European IT Conference (1994)
[1.11] Jovane, F.: Turning Manufacture into ManuFuture, Opening session. In: CIRP 53rd General Assembly, Montreal, Canada (2003)
[1.12] Jovane, F.: Research for Innovation – Rivista di Meccanica, N1061/A, Settembre (II) (1994)
[1.13] Jovane, F.: Enciclopedia Treccani, XX secolo – VII Appendice, pp. 324–328
[1.14] Sachs, W.: Transcript of the lecture on: Can globalisation become a driver for sustainable development? Wuppertal Institute (2006)
[1.15] A sustainable Europe for a better world: A European Union strategy for sustainable Development', COM, p. 264 (2001)
[1.16] Government of Japan - Science And Technology Basic Plan (March 28, 2006)
[1.17] U.S. Department of Commerce – Manufacturing In America – 2004 - A Comprehensive Strategy to Address the Challenges to U.S. Manufacturers (2004)
[1.18] China's Parliament - China's 11th Five-Year (2006-2010) Social and Economic Development Plan (March 2006)
[1.19] Article 2 of the Treaty of European Union
[1.20] European Union: Review of the EU Sustainable Development Strategy (EU SDS) – Renewed Strategy (2006)
[1.21] European Commission: Establishing a Competitiveness and Innovation Framework Programme (2007-2013) (2005)

[1.22] European Union: Progress Report on the Sustainable Development Strategy 2007 COM, p. 642 (2007)

[1.23] Measuring progress towards more sustainable Europe – Sustainable Development Indicators for Sustainable Europe – Data (1990–2005)

[1.24] Jovane, F.: ManuFuture Conference Derby (UK) (2005)

[1.25] Eurostat Structural Business Statistics - 2004 - EU 25 (2004)

[1.26] Manufuture Strategic Research Agenda (2004)

[1.27] European Parliament: A policy framework to strengthen EU manufacturing – towards a more integrated approach for industrial policy (2006/2003(INI))

[1.28] World Trade Organization: World Trade Developments – The Highlights 2007 (2007)

[1.29] European Commission – Enterprise and Industry Directorate - EU industrial structure 2007 Challenges and opportunities - 2007 EU industrial structure. "Competitiveness and Economic Reforms" (2007)

[1.30] Eurostat: Statistics in focus - R&D in Enterprises (2007)

[1.31] Eurostat: Science, technology and innovation in Europe (2006)

[1.32] 4th BMBF Forum for Sustainability: "Sustainable Neighborhood –from Leipzig to Lisbon through research", Leipzig (May 2007)

[1.33] Westkämper, E.: 4th BMBF Forum for Sustainability –Crosser Borders between Policy and Industry – Strategies to Make European Industry More Sustainable, Leipzig (May 2007)

[1.34] A New Industrial Growth. First International Forum on Sustainable Production, Venice, Palazzo Papadopoli (1996)

[1.35] Proceedings of the Second International Forum on Sustainable Production. The Role of Research, Venice (1998)

[1.36] Porter, M.: The competitive advantage of the nations (1990)

[1.37] Garelli, S.: Competitiveness Project – Competitiveness of nations: the fundamentals

[1.38] World Economic Forum: Global competitiveness report (2006-2007)

[1.39] Lopez-Claros, A., Altinger, L., Blanke, J., Drzeniek, M., Mia, I.: The Global Competitiveness Index: Identifying the Key Elements of Sustainable Growth (2007)

[1.40] Salmi, H.: Measurement of Competitiveness as the basis for policy development. European Commission Enterprise and Industry Directorate-General (2005)

[1.41] Jovane, F., Koren, Y., Boer, C.: Present and Future of Flexible Automation. Annals of the CIRP Annals 52(2), 543–560 (2003)

[1.42] Hon, K.K.B.: Performance and Evaluation of Manufacturing Systems. CIRP Annals 54(2), 675–690 (2005)

[1.43] ONU – World Commission on Environment and Development (1987)

[1.44] Elkington J.: Cannibals with Forks: The Triple Bottom Line of 21st Century Business. New society (1998)

[1.45] Seliger, G.: Product Innovation – Industrial Approach Department of Assembly Technology and Factory Management, Institute for Machine Tools and Factory Management, Technical University Berlin, Germany, Keynote paper (2001)

[1.46] The Brundtland Commission: Our Common Future (1987)

[1.47] Sustainable Development Policy and Guide for the EEA Financial Mechanism & The Norwegian Financial Mechanism (2006)

[1.48] Veleva, V., Hart, M., Greiner, T., Crumbley, C.: Indicators of sustainable production – Lowell Center for Sustainable Production (2000)

[1.49] Jovane, F.: Concepts for Strategic Machinery Innovation. ARMMS, Agile Reconfigurable Manufacturing Machinery Systems Network, Bruxelles (May 12, 1998)

[1.50] Jovane, F.: Redesigning Manufacturing: ManuFuture. In: 11th International CIRP Life Cycle Engineering Seminar Product Life Cycle – Quality Management Issues (2004)

[1.51] National Council for Advanced Manufacturing: Strategies for a new industrial Era – Sustainable Manufacturing Roundtable Report – Advanced Manufacturing Leadership forum Annual Meeting (June 7, 2006)

[1.52] NACFAM - Sustainable Manufacturing Roundtable - Advanced Manufacturing Leadership Forum Annual Meeting – Findings (June 7, 2006)

[1.53] Labuschagne, C., Brent, C.: Sustainable Project Life Cycle Management: the need to integrate life cycles in the manufacturing sector (2004)

2

Towards Competitive Sustainable Manufacturing

F. Jovane and E. Westkämper

Competitive sustainable manufacturing is the fundamental enabler of the transition from economic to sustainable development (figure 2.1). It requires transformation of manufacturing industry towards K-based, High-Value, Competitive, Sustainable Products and Services, Processes and Business Models (PS&PR&BM).

Such a transformation relies on E&RTD&I activities taking place in a subset of ERA, defined by ManuFuture as the European Manufacturing Innovation Research Area (EMIRA).

Following the previous general treatment, this chapter will deal specifically with products/services, processes, companies and business models, playing their role - within the ESET context.

Strategic objectives will be described and then, then, the existing situation and prospects of manufacturing industry will be depicted.

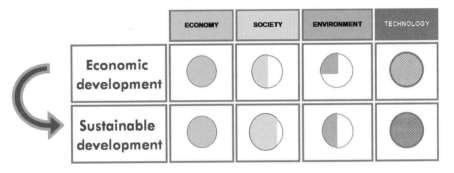

Fig. 2.1. Transition from Economic to Sustainable Development [2.1]

2.1 Competitiveness and Sustainability

2.1.1 Products and Services

Any product has an impact on the environment during its life cycle. Consumers have a significant role to play in choosing more sustainable products. Designers, manufacturers, retailers and marketing professionals should create and promote goods and services with significantly reduced environmental and social impacts [2.2].

The role of products and services is rapidly changing, due to the pursuit of competitiveness and, more and more, sustainability. A sustainable product is a good or service that complies with economy, society, environment needs and constraints. A sustainable product should minimise its impact throughout its life cycle. This may relate to raw materials, resources extraction and processing, manufacturing and suppliers, purchasing and consumption, waste and disposal, biodiversity and loss and land clearing, soil/water loss and pollution, and also exploitation and degradation, health and amenity impact [2.3].

The environmental impact of a product may be different, qualitatively and quantitatively, at each stage of its life cycle. This will affect its overall degree of sustainability. Finally, products and services must be considered jointly with processes and business models concerning their life cycles.

Competitive Sustainable Manufacturing (CSM) aims at a new industrial revolution [2.4], where preservation of natural resources reaches a level of importance comparable to labour and capital assets and the contradiction between wealth creation and raw material preservation may be overcome by *knowledge* becoming a productivity driving factor that allows the possibility of value-creation patterns based on intangibles.

In the resulting paradigm [2.4], the basic proposal for environmental sustainability shifts the market, from the trade of artefacts, to that of intelligent products services capable of equivalent functions, in terms of consumer satisfaction. Such a market would be supported by intelligence suppliers, with profitability based on knowledge intensity, rather than on material content. In terms of quality, we move towards HAV Sustainable Quality (SQ), as shown in figure 2.2 [2.5].

T. Tomiyama [2.6] proposed the Post Mass Production Paradigm (PMPP) as a system of economic activity, capable of encouraging and sustaining economic growth without depending on mass production and mass consumption of artefacts. PMPP may be seen as a way of decoupling economical growth from resource /energy consumption and waste creation thus pursuing global sustainability.

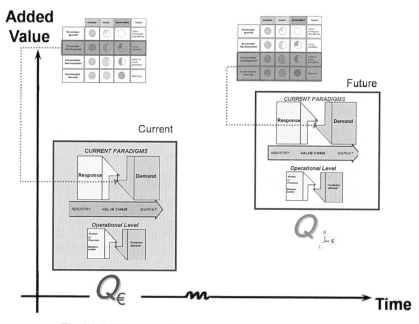

Fig. 2.2. Moving towards HAV Sustainable Quality (SQ) [2.5]

PMPP requires the development of new kind of products, called "soft artefacts". They should overcome current problems and compensate decrease in production volume through the added-value generation with accumulated intensive knowledge in life cycle product.

Yoshikawa [2.7] states that what people value is not a product itself, but its functionality. This is the service embedded in the product. Latent functionality appears as service when the product is used. Functionality of a product decreases when it is used. Therefore, we can measure the potential value of a product by functionality – that is the total amount of service available. As for the related production process, Yoshikawa states that sustainability requires integration of manufacturing and inverse manufacturing, i.e. closed loop manufacturing, covering from construction to implementation, extraction and dissolution of functionality. Within this frame, minimal manufacturing and maximal servicing paradigm for sustainability produce the product value.

Innovation [2.2] is needed to improve products and services, as well as processes, business models and organisations, to create new, less resource-intense ways of meeting the needs of customers. Thus, innovation in the design of products and services can lead to the creation of new markets, promote competitiveness and enhance corporate reputation, at the same time as delivering social and environmental benefits.

Sustainable innovation is mandatory, as sustainability considerations (environmental, social and ethical, financial) must be integrated into company systems, from idea generation through to research and development and commercialisation. This applies to products, services and technologies, as well as new business and organisation models [2.8]. Sustainable innovation is paired by eco-innovation, aiming at reducing impacts on the environment or achieving a more efficient and responsible use of natural resources, including energy [2.9].

Product innovation, see par. 4.4, requires method and function innovation [2.2]. Method innovation should provide structured procedures to identify non-renewable resource balance and appropriate solutions, empowering intelligent enterprises to control the consumption of natural resources. It would include tools and methodologies encompassing from environmental to societal and economic analysis and design, concerning products and services, and related processes and business models. Function innovation would aim at supplying functions rather than tangibles and achieve an equivalent end result. This would be helped by consumers' behaviours along different targets, from replacement, to equivalent satisfaction, fair trade, accreditation.

Solving product and service issues may lead to relevant effective benefits, as compared to the traditional manufacture market. Extended enterprise capabilities could grow by information sharing with actors in the extended-product chain. The scenarios open challenging prospects, once proper scientific and technical know-how is developed and shared.

The current structure of industry and the considerable amount of international trade requires a strategy that takes these conditions into account in advancing sustainable production and consumption. The European Union has developed an Integrated Product Policy (IPP), built on life cycle thinking, in order to reduce resource use. The aim of IPP [2.10] is to minimise the impact of products on human health and the environment throughout their life cycles, from cradle to grave, to improve sustainable production and consumption, and advance the government's environmental quality objectives [2.11]. At the same time, it also involves many different actors such as designers, industry, marketing people, retailers and consumers. IPP attempts to stimulate each part of these individual phases to improve their environmental performance.

In order to implement an IPP approach forthcoming actions are strengthening green public procurement, revising the EMAS and eco-label schemes and fostering research into less resource intensive products and production processes. [2.12]

Governments worldwide have called for increased international co-operation on policy to encourage more environmentally sustainable, energy efficient products. The International Task Force for Sustainable Products (ITFSP) [2.13] has been established to respond to these needs. Led by the United Kingdom, with participants from 13 countries, the IEA and the UN, the ITFSP aims to identify the priorities for action, then stimulate and support the development of international networks and co-operative projects to address the key priorities.

2.1.2 Processes

Production can be seen as a transformation process (figure 2.3) in which employees use machines and energy to transform material into products for customers use. In the production process losses of material and energy generate emissions and waste. An indicator of the effectiveness of the transformation process is the efficiency of all resources.

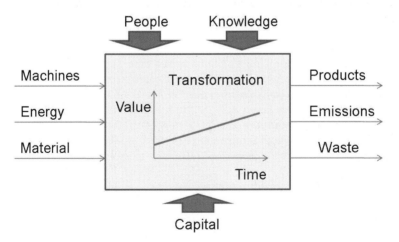

Fig. 2.3. Production as Transformation Process ©IFF/IPA [2.14]

The economic efficiency – measured by the relation of the costs of resources to the value of products – is the main criteria for competition. Increasing costs of inputs, which are caused by world market conditions for energy and material and the local costs of labour and the inefficiency of processes, reduce the Adding-Value and the profitability of capital.

Understanding the transformation as a complex socio-economic and ecologic system for generating value, manufacturing has to take the sustainability as a major criteria and strategic objective for High-Adding-Value.

In practice the transformation process follows a complex chain and division of labour. Metal and polymers are the main material sources. Parts and devices including electronics and components are the elements of today's technical products. They are assembled in the final stages to more or less complex products. Manufacturing consumes an increasing amount of natural resources. The integration of electronics into products turned them into mechatronic systems.

The markets changed from mass production to a higher variety and diversity towards customisation. Shorter life time, shorter time-to-market and higher functionalities are consequences of the technical development of products and market requirements. Mass production followed the migration to regions with lower costs of manufacturing. The structural change of manufacturing industries – driven by competition and technical innovations – is still going on.

Discussions about the circular-flow economy opened the way to the life cycle of products (figure 2.4), in which the phases of usage and "de-production", became elements of the value chain. Product-oriented services, disassembly, remanufacturing and recycling are manufacturing processes for retrieval and to gain environment's value.

Europe made big steps forward to the circular-flow-economy. Regulations for emissions during the production, usage and de-production phases in the life of products and the integration of knowledge about hazardous materials or fluids pushed the development of manufacturing. Today's European factories are clean and have a high ecological standard.

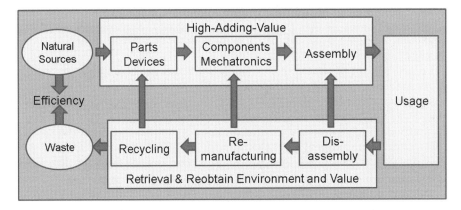

Fig. 2.4. Manufacturing in the Circular Economy ©IFF/IPA [2.15]

Even new fields of industrial activities like disassembly, remanufacturing and recycling are sources of Adding-Value. Last but not least, the social conditions of work follow the European culture and regulations between capital and work (figure 2.3). But the efficiency of the main resources (material and energy) is still disappointing. Even the division of labour in the chains creates losses of the economic efficiency. High division of labour and the customisation in manufacturing causes turbulences and losses of the overall economic, ecologic and social effectiveness in the life of products. The main objective is the maximisation of effectiveness of each technical product over the life cycle. This objective can be summarised as *sustainable manufacturing*.

Sustainability relates to the continuity of economic, social, institutional and environmental aspects of the human society as well as the non-human environment. The environmental aspects get high priorities, when we accept the threats of global warming, climate scenarios and the consumption of natural resources. The responsibility for maintaining the economic base, the impact on the environment and the social/cultural influence of products made by manufacturers require change of paradigms and objectives in manufacturing. All experts expect increasing costs of input and costs of disposal of waste. The European Commission and the governments intensify the social and environmental regulations. This creates additional economic inefficiencies when enterprises only follow the pressure of surrounding conditions. Manufacturing in Europe has the potential for preventive activities and proactive Research and Technology Development (RTD) for the integration of economy, society and ecology.

2.1.3 Companies and Business Models (Integration of Sustainability Aspects in Production Systems)

The economic effectiveness includes the survival of enterprises in turbulent environments and under the dynamic influencing factors. Short term profit optimisation is not the answer for sustainability. But enterprises need the profitability to survive and to transform the structure of the socio-economic/ecologic system for High-Adding-Value. The whole field of methodologies and the potential of old and new technologies can play a strategic role in all sectors of the manufacturing industries. By integration of economic, ecologic and social aspects the effectiveness of future manufacturing can be increased to a leading level (figure 1.16). Just European research has the potential to integrate the different aspects and join them in the factories and management systems of enterprises (figure 2.5).

The management system needs methodologies which can be implemented in a holistic production system with a European "brand".

The European quality level is quite high. Sustainable quality refers to long-life products and low consumption of energy and material.

The efficiency of processes has a strong relation to the effectiveness of the manufacturing system. There are many examples of initiatives, which seemed to be a cost-driver but we know their impact on system effectiveness. One example is the implementation of Total Quality Management (TQM) in the overall production system.

Quality management followed consequently customer-orientation and proactive reduction of defects. Thus, the supply chains in production networks became much more effective and many enterprises earned higher profitability and competitiveness.

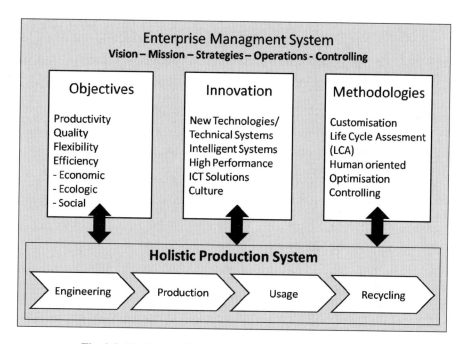

Fig. 2.5. Challenges of Sustainable Enterprises ©IFF/IPA [2.16]

Similar initiatives are necessary to reduce the losses of energy and material in manufacturing and to reduce the pollution of emissions (water, air, noise, etc.). Input and output factors and losses in the chain from original product manufacturing to the end of life of products have a high potential for efficient processes and contributions to economic competition.

2.1.4 Strategic Objectives

Manufacturing as a part of society and the European culture is influenced by economic, social and political interests as well as by influences of future mainstreams as environmental impacts. Following the Lisbon Agenda and the new agenda for sustainability, defined by experts in the Conference of Leipzig 2007, strategic orientations have to be found. Of course economic growth, competition and high employment are known requirements of the European policy for Adding-Value. High-Adding-Value is required for welfare and the economic base of living.

The problems of environmental impacts increase and it is now necessary to add sustainability as a strategic objective. Sustainability includes the sustainability of businesses (capital efficiency) and the orientation towards efficiency of material and energy, used for the manufacturing of technical products. The impacts of emissions on the climate are known and they create new threats but even a potential of future manufacturing.

The pursuit of technological and organisational transformation, described in the ManuFuture Vision document, will be crucial for the future success of European

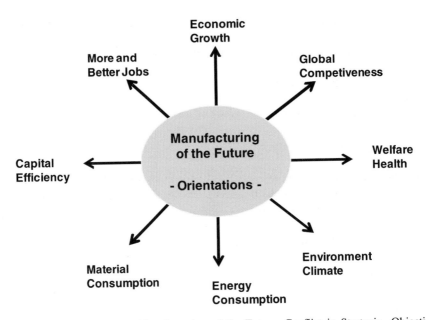

Fig. 2.6. Orientation towards Manufacturing of the Future: Conflict in Strategic Objectives ©IFF/IPA [2.17]

manufacturing in sustaining and strengthening this vital activity. The orientation towards manufacturing of the future is presented in figure 2.6.

The activities recommended in the following pages focus on the realisation of a set of four strategic objectives:

- Competitiveness of sustainable European manufacturing industries:
  - to survive in a turbulent economic environment by obtaining sustainable economic results and transforming them into a European improvement of quality of life,
  - to benefit from the migration of technologies,
  - to create more and High-Added-Value jobs,
  - to install a culture of continuous innovation in products and services, concentrating on innovation efforts on cross-sectorially important technologies.
- Leadership in manufacturing technologies:
  - to support innovative products and platforms,
  - to lead manufacturing with global standards,
  - to reach the forefront in networking,
  - to ensure competitive efficiency and labour productivity.
- Eco-efficient products and manufacturing:
  - to reduce adverse environmental impacts,
  - to cut the consumption of limited resources,
  - use of renewable resources,
  - to maximise the benefits of each product throughout its life cycle.

- European leadership in cultural, ethical and social values and also in processes and products:
  - to understand the interrelationship between social and ethical values and the prosperity of an enterprise,
  - to ensure welfare and social standards of living, supportive of innovative and entrepreneurial businesses,
  - to guarantee human and social standards of work.

2.1.5 Current and Perspective Challenges

Understanding manufacturing as a process of transformation of input to output – the machine for Adding-Value – high effectiveness supports even competition in the global economy. Waste caused by loss of energy and material and inefficient processes reduces the Added-Value. Even losses caused by defects and inefficient work (labour) contribute to lower economic benefits. In the past three groups of actors: manufacturers of products, users and recyclers optimised the benefits in the life of each technical product.

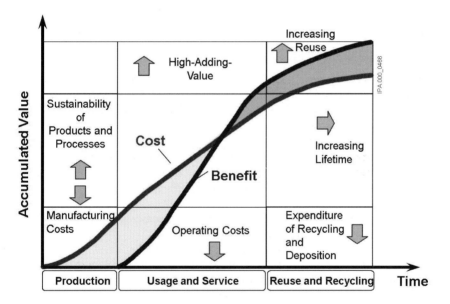

Fig. 2.7. Costs and Benefits in the Life Cycle of Products ©IFF/IPA [2.18]

In the future and referring to the needs of sustainability it is of fundamental interest to optimise the benefits of each product in its perspective life cycle.

Figure 2.7 illustrates the lines of costs and benefits in the life of products. Considering the increasing costs of energy and material it is necessary to increase the efficiency of the production system and change traditional ways of cost-optimisation towards a life cycle paradigm.

The drivers for a higher efficiency are technologies of products, processes and the management of manufacturing. The change towards the life cycle paradigm can be realised in nearly all sectors of industries through:

- methodologies for life cycle design and assessment,
  holistic production system with criteria of life cycle management
- technologies for increasing the technical efficiency of processes and products,
- sustainability standards,
- machines and systems for low energy consumption,
- emerging areas of new technologies for sustainable products and services.

European manufacturing has the competence for the implementation of sustainable manufacturing and is able to lead the world markets. There are many successful industrial examples which show the effect of increasing effectiveness on the profitability and Adding-Value. Industries give the research for reducing energy and material high priorities and grow in areas like solar-technology, fuel cells or others where alternative renewable resources can substitute fossil material.

2.2   Manufacturing Industry: Existing Situation and Prospects

The manufacturing industries in Europe are fighting for competition because of high wages and costs of resources. Companies learned to concentrate operations towards customisation and niches of higher profitability. Taking into account the high skill of workers and engineers customisation changes the structure of manufacturing:

- increasing variants and customer specific products,
- lower batches and resulting costs for transformation,
- increasing complexity of products,
- increasing costs of product development.

The transformation from mass production to customised production is the new challenge of manufacturing.

2.2.1   Manufacturing in Europe

European manufacturing industries are the backbone of the economy and welfare of European people (figure 2.8). Their good position in the world markets is attacked by other economies and by the key challenges, as reported in paragraph 1.2.2. All the main data concerning manufacturing and related E&RT&I system are presented in paragraph 1.2.3.

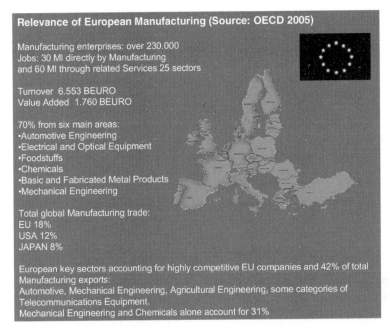

Relevance of European Manufacturing (Source: OECD 2005)

Manufacturing enterprises: over 230.000
Jobs: 30 Ml directly by Manufacturing
and 60 Ml through related Services 25 sectors

Turnover 6.553 BEURO
Value Added 1.760 BEURO

70% from six main areas:
•Automotive Engineering
•Electrical and Optical Equipment
•Foodstuffs
•Chemicals
•Basic and Fabricated Metal Products
•Mechanical Engineering

Total global Manufacturing trade:
EU 18%
USA 12%
JAPAN 8%

European key sectors accounting for highly competitive EU companies and 42% of total
Manufacturing exports:
Automotive, Mechanical Engineering, Agricultural Engineering, some categories of
Telecommunications Equipment.
Mechanical Engineering and Chemicals alone account for 31%

Fig. 2.8. Manufacturing in Europe [2.19]

2.2.2 Changes of the Global Markets

The globalisation of the economies has chance and risks. Up to now, the countries of the Triade (US, Japan, Europe) were leading the world markets and technologies. Economic conditions of manufacturing of these regions are in a level of acceptable conditions. They are the leaders of technical standards and main investors in the world.

The leading position of European manufacturing is now threatened by the economic developments in Asia. China and India have a growth of their GDP about 10% yearly. China is on the way to lead the world markets again and remember their position hundreds of years ago. High population is the base of immense human capital and changes in the economic system from centralised to decentralised structure, deregulation of former policies (China) and privatisation of enterprises created a climate of growth (figure 2.9).

The social and environmental standards are still behind the Triade but the competition learns fast and has opened the human potential and skill towards the global technical standards. At the same time, the growing economies in China and India are growing markets for European manufacturers and cheap resources for engineering and supply. Plagiarism is rising to close on the world state of the art in technologies and especially in manufacturing know-how.

Growing economies create increasing welfare but even the environmental impact rise up now to a level of limitations. The availability of energy and material let costs worldwide explode. Waste and emissions of industrial manufacturing and usage of products intensify the problems of global environment and cause disadvantages for the traditional industries.

Today's growing markets in the economies of China, India and some others are opportunities for the expanding sales of European industries by activating the low-cost potential of these countries and by the fact of their high population.

Both countries developed their technological potential and have high advantages in manufacturing costs by low wages and working conditions. They are fast followers in the technological competitiveness and productivity by implementation and of course imitation of knowledge. The differences of labour costs are in the dimension of factor 10, as shown in figure 2.10. Experts expect increasing costs of labour in China and India but even in ten years the differences will still be high.

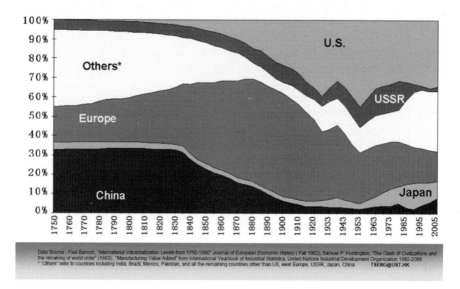

Fig. 2.9. Changing the Economies of the World [2.20]

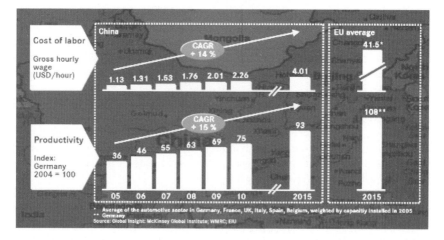

Fig. 2.10. Development of the Costs of Labour and Productivity, Europe versus China © McKinsey Global Institute [2.21]

The level of productivity in China was only 36% of Germany's productivity. But China's rate of development is high. Main factors are the global availability of know-how and the implementation of world class technology in the capacities for manufacturing. It is expected that China will almost have the German level of productivity within the next decade.

2.2.3 Migration of Production and Consumption

Changes of economic conditions in the global age, caused by open markets and technologies, increase the speed of migration, consumption and production of technical products. The global mobility of goods, money, people (business) and information has been accelerated by world standards and regulations between countries. Manufacturing seeks growing markets and profitable economic conditions. The migration process is not new – known even in the industrial age with revolutionary structural changes for economies and people (figure 2.11).

The speed of migration has been accelerated and causes unemployment in Europe as well as the globalisation of operations. Factors of acceleration of the migration of markets are:

- costs of resources (labour, material, energy),
- open markets and global logistics,
- transfer of technological knowledge,
- open and global systems of communication,
- changing political and economic systems.

Of course the economies in the Triade (Europe, USA, Japan) have a long tradition in social and cultural aspects of work: social and cultural standards, regulations of

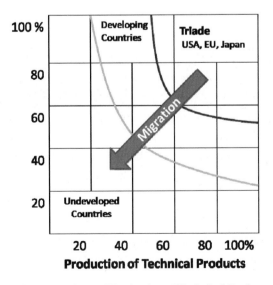

Fig. 2.11. Migration of Consumption and Production of Technical Products   ©IFF/IPA [2.22]

waging work safety and ergonomics, unions of workers, environmental regulations. These standards are lower or partly not yet implemented in developing countries.

China and India have the highest growth of economies and strong expanding market volumes. Manufacturing follows markets to achieve advantages in market positions and market orientation. Experts expect growing markets in China and India not only in low but even in the high-tech sectors, competing against Europe's, the US's and Japan's leading industries.

Mass production and low technologies migrate to the developing regions. This causes problems in European regions (East and West) with lower labour costs than in central Europe. Sectors of consumer goods like textile, shoes, consumer electronics are flowing to the developing countries. Far East manufacturers of electronic components are nearly completely leaders of the world markets.

The low technology sectors of industries are not in the centre of strategic research and the European policy underestimates the economic and technical potential. There are some very old technologies, which are required for future products as well as for the efficiency of the technical base of industries. Many of them have a continuous development, broad and trans-sectorial relevance, but they are undervalued in research strategies.

Growth of the economies requires competitive manufacturing in the political, economic and ecologic environment. Changes of the conventional and mainly profit oriented business objectives are necessary and this will change the economic system in Europe and create new market potential for High-Adding-Value.

2.2.4 Economic Potential of Manufacturing

Manufacturing is a trans-sectorial area of the European economy, with a high variety of products. They range from mass production of consumer goods to special solutions (unique). The technological processes include traditional as well as high's innovative technologies and services along the life cycle of products. Big and globally operating Original Equipment Manufacturers (OEMs) and many small and medium enterprises illustrate the variance of the structure.

Networking and Supply Chains characterise the way of today's manufacturing. Under the conditions of regional or global markets towards market-orientation and customisation the structure changed. Time-to-market and the efficiency of networking in manufacturing are critical factors of success. Europe has a highly developed infrastructure for the networking and engineering of products and even a high potential of productivity.

Europe's innovations in industrial technologies made sectors successful: e.g. automotive, electrics, machines, aerospace. Some of them are concentrated in regions, which host the leaders of the world markets. Europe generated its own ways of manufacturing or adapting profitable solutions from the U.S. or Japan. Mechanisation, automation, Taylorism, electronics in manufacturing and ICT systems for manufacturing are some of the solutions with high contributions to the growth of productivity and the competitiveness of European industries. It made Europe and especially some regions into high wage regions in which people benefitted from the economic success.

Competitiveness depends not only on the level of wages but even on the level of innovation in technologies and methodologies and the efficiency of networking in manufacturing. European manufacturers have a strong part of the global market and some of the European sectors are world market leaders.

As anticipated in paragraph 2.1.4 to make this position "sustainable" and to grow, a strategic push is necessary in all industrial sectors. Restructuring towards future manufacturing is essential.

Compared with other fields of research, research for manufacturing has to be oriented to innovation and rapid implementation to get advantages in the global market. Long term industrial perspectives need strategies from survival in turbulent economies up to long term implementation of future fields.

2.2.5 Industrial Structure

The industrial manufacturing enterprises can be divided into two groups:

- manufacturers of capital intensive goods and services, who produce and deliver material, component machines, systems and equipment for manufacturing and
- manufacturers, who produce products for consumer markets.

Both groups need manufacturing technologies for their production. The productivity of manufacturing depends critically on innovations and efficiency of solutions, made by the manufacturers of capital intensive products. The group of manufacturers of capital intensive (investment) goods have a dominant role in the development of technologies and machines for manufacturing. Innovations in the capital intensive sectors boost the competiveness of all industries. Investment goods are long-life products. Their capability influences the productivity, flexibility and the structure of manufacturing over a long time (10-15 years). The technical solutions have influence on the quality, efficiency of material and energy and on the conditions of work (human–machine).

The level of productivity of all industrial sectors depends on the state of the art of the technology implemented in machines and technical systems: automation. These industries are the enablers of manufacturing of the future. It is of a strategic dimension to lead the world market in these fields. The market potential is in dimensions of some 1000 BEURO.

European manufacturers of capital intensive products are technological leaders. This is based on:

- strong customer orientation of engineering and specific solutions,
- high qualification of workers and technicians,
- networking and co-operation,
- infrastructure of logistics and communication.

But they are weak in manufacturing of mass products and standards. Many manufacturers have lost markets of series products under the conditions of high wages and European surrounding conditions.

The sectors of consumer goods are manifold. They range from big and capital intensive products like airplanes, ships, and trains over medium capital intensive technical goods like automobiles and go down to mass products for people like textiles, sports or other goods. They even lost markets of mass products under the economic conditions and oriented their strategies to higher variants and market niches (figure 2.12).

Manufacturing, customisation gets more and more relevance in the future. Main drivers are of course the profitability and the technical mainstreams towards configurable product families. Customisation reduces the lot sizes and increases the costs of adaptation and management.

All industrial sectors have strong competitive disadvantages and are resorting to specific core competences. The results are higher skills and lower employment. Loosing mass markets and markets for lower technologies the unemployment, especially of lower skilled people, will increase.

The proportion of engineers according to the number of employed people in German Machine Industries was at 8% in 1995. It is now at 16% and will come up to 20% in the next five years.

This structural change is mainly caused by losing markets of series production and low technologies (measured by the relation of RTD to turnover). 60% of the German manufacturers spend less than 1% of their turnover for RTD and nearly nothing for RTD for manufacturing. They are extremely vulnerable to survive in the global migration process.

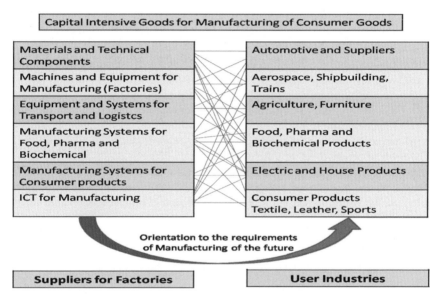

Fig. 2.12. Industrial Sectors of Manufacturing ©IFF/IPA [2.23]

There is another structural change in the manufacturing industries. Caused by the requirements of profitability, nearly all manufacturers are increasingly reducing their in-house production to the core competences by outsourcing. The efficiency of the resulting networked production depends on the efficiency of suppliers and the logistic system. The trend towards more customisation and protection of knowledge causes structural changes in the future.

SMEs are the backbone of the capital intensive goods sectors. Their diversity and flexibility as well as regional engagement stabilised the European industry in all sectors, but they are influenced by cyclic markets. Many SMEs with world class manufacturing disappeared in times of economic downturn and low investment rates. They are extremely vulnerable in traditional sectors of low technologies. Research for survival in turbulent markets and activating the potential of low (traditional) technologies is essential for Adding-Value and jobs.

2.2.6 European Strengths and Weaknesses

A number of European strengths and weaknesses can be identified:

Strengths

- European industry is modern and competitive in many areas. A long-lasting industrial culture exists, with large networks linking suppliers, manufacturers, services and user companies,
- high level of skilled workers,
- leading-edge research capabilities are available across member states, leading to high levels of knowledge generation and a reputation for scientific excellence,
- some 99% of European businesses are SMEs, which typically exhibit greater flexibility, agility, innovative spirit and entrepreneurship than more monolithic organisations. In addition, SMEs tend to interact in a manner that lies between strong competition and fruitful co-operation, which helps to foster the process of what has been called 'co-opetition',
- Europe has taken on board sustainable development. Significant investments in environmental protection, clean technologies and environment-friendly production processes have led to new manufacturing and consumption paradigms,
- historic and cultural differences between individual member states and regions bring a diversity of viewpoints and skills that can be co-ordinated to produce novel solutions.

Weaknesses

- Productivity growth in European manufacturing industry as a whole has been below US levels in recent years. Investment in ICT and new technologies is still too low, and has not so far led to the desired productivity gains; and
- innovation activity is too weak. The EU does not suffer from a lack of new ideas, but is not so good at transforming these into new products and processes. Industry's analysis is that this is due to the framework conditions for manufacturers operating in Europe,
- Europe has a dislocated infrastructure in research,
- multiple models of research for application.

2.3 Leadership in Manufacturing

The European industries for capital intensive products and equipment for manufacturing have a position of 22% in the world markets. Some of the sectors are leading in technologies and quality but on a high level of costs. Some produce with low and

**Assuring the Future of Manufacturing in Europe
by High Adding Value**

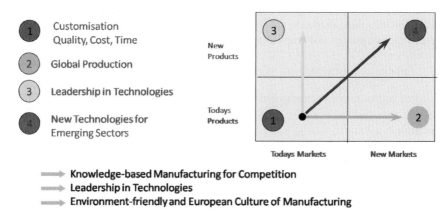

1   Customisation
    Quality, Cost, Time

2   Global Production

3   Leadership in Technologies

4   New Technologies for
    Emerging Sectors

⟹ **Knowledge-based Manufacturing for Competition**
⟹ **Leadership in Technologies**
⟹ **Environment-friendly and European Culture of Manufacturing**

Fig. 2.13. Portfolio and Strategic Goals of the Manufacturing Industries ©IFF/IPA [2.24]

conventional technologies under extreme cost pressures. There are four fields in the portfolio, which characterise the strategic goals of the development (figure 2.13) by strengthening their position.

Following the objective of High-Adding-Value, European manufacturing industries are able to lead markets in traditional sectors like machines and systems for production as well as in emerging sectors like the healthcare technology sector or energy systems. They have the necessary engineering potential for products of the future with a wide spread competence and the infrastructure for an efficient production but competitive disadvantages by high wages and high costs of the social system.

High-Adding-Value, based on continuous innovation and efficiency, seems to be the only way to competitiveness in the global markets.

2.3.1 Customisation and Life Cycle Orientation

Customised Products are the specialty of traditional manufactures of machines and systems. Under the surrounding conditions of markets and economies European manufacturers lost market shares in the mass production areas. Quality and efficiency in cost and time made European manufacturers leaders in many sectors with specialised solutions and high levels of reliability. Machines made in Europe are costly but efficient in the life cycle.

The vision of the future is characterised by extreme customisations and customer orientation in all operations in the life cycle of products from customer requests to the end of life. Customers are in the centre of management and product engineering, manufacturing and after sales activities. The position of European manufacturing will be boosted by life cycle orientation, life cycle management and activation of value in product-related services. Innovative solutions for life cycle management, new business models and global ICT services are required. The idea of life cycle – machines and technical products remain in the network of manufacturers – opens a wide field of Adding-Value and implementation of knowledge as part of the knowledge economy.

Life Cycle Assessment (LCA) takes into account the impact of products and production on the environment. Life cycle management activates the Adding-Value of products from engineering to the end-of-life with the objective of maximising the economic benefit at the end of products.

2.3.2 Global Production

European manufacturers have a world market share of round about 20%. They are familiar with all markets. But, most of the manufacturers are SMEs which don't have the financial and human potential to realise the chances of global markets or to fight against plagiarism (paragraphs 1.2.3, 2.2.2). According to the CSM emerging paradigm global prosperity depends on economic policy, regulations and open markets.

Research can be contributed by investigations in market conditions, market analysis, risk management, resource management and IP management. Additionally, fairs and marketing strategies are action fields for supporting the global position and winning of market shares. Expansion towards a market share of more than 50% seems to be possible by fastening the innovations and the diffusion of innovative technologies into industrial practice. Time is one of the critical success factors for global manufacturing and it is known, that technological leaders, who cover the global markets first, can reach leading positions in the global market. There is a chance for market domination by excellence in technologies, customer orientation and innovative production in nearly all sectors of manufacturing. The domination of the way of manufacturing, based on excellent factories "Made in Europe", should follow the vision shown in figure 2.14.

The vision of ManuFuture is leading manufacturing to factories, which are based on a high efficiency and economy and on European manufacturing standards for

Fig. 2.14. World Supplier of Factories and Equipment ©IFF/IPA [2.25]

global markets. European manufacturing has a long tradition and culture of work in factories and a high competence in technology. European manufacturing industries have learned permanent innovation and customisation. They have the potential for making standards in technologies, in management systems, in social and human conditions and in environmental solutions. But manufacturers have to push innovation by fast implementation of basic knowledge.

2.3.3 Leadership in Technologies

European manufacturing is driven by engineers. It is typical for engineers to create technical solutions for customer demands and overcome existing limits of the technologies state of the art. Based on high skills of engineers, technicians and workers in Europe and on competences in all future basics it seems to be possible to innovate permanently by overcoming existing technological limits and to activate the technical potential (figure 2.14).

Some of the manufacturing technologies like cutting have a long tradition but they are not at the end of capability. Technological developments were driven by innovations for a higher productivity and profitability towards higher performances. Changing factors of manufacturing and basic research open a new area for future manufacturing by activating technological potential in clean manufacturing, low energy consumption and emission of pollution, lowest consumption of material or intelligent generative processes. Industrial practice and research are far away from physical or chemical limitations. The knowledge generated by research can push innovative technical solutions permanently, even in low technologies and open possibilities for future innovative products.

Especially the consumption of energy in manufacturing has a high priority as a consequence of the growing increase of costs and the environmental priorities (climate). This topic includes technical solutions for lowering technical losses and efficiency of the energy supply chains in manufacturing. Factories are complex systems for the manufacturing of products.

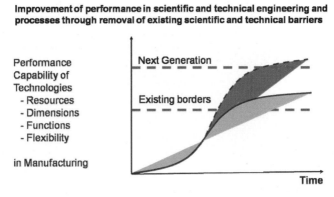

Fig. 2.15. Overcome Existing Technical Limits for Leadership in Manufacturing ©IFF/IPA [2.26]

The fields shown in figure 2.15 are actual topics to develop and implement technologies of the future in factories. They include technologies which enable users to manufacture as efficient as possible.

But the time for implementation and increase of the state-of-the-art in industries is the critical success factor (Time-to-market). Companies, who are behind, lose market shares. Research has to contribute to the process of implementation by a strong orientation to application and fast diffusion of results and best practices in the wide field of manufacturing industries. Achieving the leadership in technology demands a long term strategy and includes basic research. Experimental-based research and development is cost and time intensive. In the long-term modelling and simulation, enabling technologies are for the forecast of the behaviour and the impacts of new technical solutions in the life of each product. They are the enablers of knowledge-based fast engineering.

2.3.4 New Technologies for Emerging Sectors

Customisation and product life cycle management, leadership in technologies and actions to strengthen the global prosperity are the main strategic options for Adding-Value and competitiveness of the manufacturing industries. The sectors of capital intensive production are the enablers for the future of manufacturing. Increasing their world market share from today's 20% to over 50% and making the world standards of manufacturing is the highest push for European economies.

Besides this, there is even a broad field for emerging sectors, whose development depends on manufacturing technologies. For the main sectors like:

- technologies for health sectors (e.g. medicine),
- life sciences (bio-based technologies like manufacturing of pharmaceutical),
- renewable energy systems (solar technologies, bio fuels).
- environmental technologies (catalyser),
- entertainment products.

Manufacturing can take the role as industrial enabler by knowledge-based engineering, intelligent production process technologies, machines and systems for manufacturing and management methodologies. It is essential to see that the industrialisation and reproducibility as well as the speed of application depend on the availability of industrial solutions for manufacturing. Some of the emerging sectors promise big markets and require specialised factories.

References

[2.1] Jovane, F.: How to Maintain Competitiveness on the way to Sustainable Manufacturing – Tampere Manufacturing (summit, June 2007-2008)
[2.2] UK government: Key priority areas for sustainable development – Sustainable production and consumption: How to create better products and services –sustainable-development.gov.uk
[2.3] NSW Government – Sustainable Procurement Programme - Purchasing sustainable products - What is a sustainable product? (2008)

[2.4]   Michelini, R.C., Razzoli, R.P.: Product-service eco-design: Knowledge-based infrastructures. Journal of Cleaner Production 12 (2004)
[2.5]   Jovane, F.: How to Maintain Competitiveness on the way towards Sustainable Manufacturing – Tampere Manufacturing (summit, June 6-7, 2007)
[2.6]   Tomiyama, T.: A Manufacturing Paradigm Towards the 21st Century. John Wiley and Sons, Chichester (1997)
[2.7]   Yoshikawa, H.: Full Research for Sustainable Industry National Research Institute of Advanced Industrial Science and Technology NTVA Technology Forum, September 26, Trondheim (2006)
[2.8]   Charter, M., Clark, T.: Sustainable Innovation - Key conclusions from Sustainable Innovation - Conferences 2003–2006 organised by The Centre for Sustainable Design - The Centre for Sustainable Design (May 2007)
[2.9]   Competitiveness and Innovation Framework (2007–2013)
[2.10] European Commission: Communication from the Commission to the Council and the European Parliament – Integrated Product Policy Building on Environmental Life-Cycle Thinking (2003)
[2.11] Reinhard, Y., et al.: Towards Sustainable Products in Sweden. Swedish Environmental Protection Agency Environmental Progress (2003)
[2.12] Commission of the European Communities: Communication from the Commission to the Council and the European Parliament. Progress Report on the Sustainable Development Strategy 2007 – COM 642 final. Brussels (2007)
[2.13] http://www.itfsp.org/aims.htm
[2.14–2.18] Westkämper, E.: Private Archive
[2.19] Jovane, F.: ManuFuture Platform, TUBITAK, Ankara (2006)
[2.20] Mitchell Tseng – Advanced Manufacturing Institute - Hong Kong University of Science and Technology (2007)
[2.21] McKinsey Global Institute
[2.22–2.25] Westkämper, E.: Private Archive
[2.26] Westkämper, E.: Manufuture Strategic Research Agenda (2004)

3

The European Strategic Initiative ManuFuture

F. Jovane and E. Westkämper

The *European Technological and Industrial Revolution for global competitiveness and sustainability* of manufacturing is fundamental for European industry. It calls for a structured approach, encompassing – within a robust reference model for action – from SI, to a framework for SI implementation, to stakeholder roles, to resources involved, focusing on transformation of industry and E&RTD&I, to pursue CSM. ManuFuture is a European response to the aforementioned needs.

In this chapter, the European ManuFuture initiative is described, covering from the ManuFuture Platform, to Vision 2020, SRA features: i.e. K-based manufacturing and Roadmap for industrial as well as E&RTD&I system transformation, drivers of change, pillars and domains of actions, multi-level action. Further, this chapter reports on the current situation of the education and RTD&I system in Europe and perspective transformation required as emerging from the SRA. Issues concerning investments in RTD are dealt with.

3.1 The European ManuFuture Initiative

Public institutions in advanced countries, such as Japan, USA and Europe, are more and more concerned about manufacturing's future, due to its socio-economic and technological impact. Emerging countries, such as China, are devoting increasing attention to manufacturing policies.

As anticipated in paragraph 1.2, Japan's government released its Third Science & Technology Basic Plan to cover policies during 2006-2010 [3.1].

China has set out her development policy guidelines in the 2006 Five Year Plan which – while acknowledging that economic reform, growth, and development will continue at a high rate – now places emphasis on the development of a harmonious society in which more consideration is given to the social implications that are associated with rapid economic development [3.2].

In the US the Department of Commerce launched the "Manufacturing Initiative" aiming to develop a strategy designed to foster U.S. competitiveness in manufacturing and stronger economic growth at home and abroad [3.3]. In order to offer effective and continued support to U.S. industry in its sustainable manufacturing efforts, the Department of Commerce is carrying out a public-private dialogue that aims to identify U.S. industry's most pressing sustainable manufacturing challenges and ways for the public and private sector to work together in addressing such challenges.

In 2003 the Industrial Technologies Directorate of the European Commission's Research DG launched the – now autonomous – ManuFuture initiative. Its mission, as anticipated, is to pursue HAV, K-based, CSM, involving stakeholders, from policymakers, to public authorities and financial institutions, to industry, university, research institutes and centres.

To this end, ManuFuture has developed and is proposing and implementing a strategy, based on research and innovation, aiming at CSM, capable of contributing to increasing the rate of industrial transformation in Europe, secure HAV employment and win a major share of the world's manufacturing output in the future knowledge-driven economy. Figure 3.1 shows the ManuFuture initiative in progress.

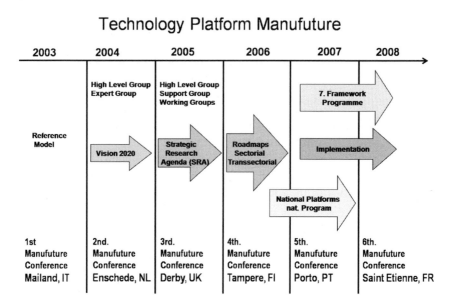

Fig. 3.1. The ManuFuture Initiative in progress ©IFF/IPA

The ManuFuture Initiative's strategic objectives are:

- competitiveness of sustainable European manufacturing industries,
- leadership in manufacturing technologies,
- environment-friendly products and manufacturing,
- European leadership in cultural, ethical and social values.

A first study, see figure 3.1, was carried out by expert groups, set up by the Industrial Technologies Directorate of the European Commission's Research DG. The document [3.4] was discussed and approved by the first ManuFuture Conference [3.5], held in Milano in December 2003. The ManuFuture 2003 Conference confirmed the need for:

- a changing paradigm, moving towards sustainability and competitiveness, based on HAV, K-based products and services, processes, business models (CSM),
- transforming industry and its sustaining E&RTS&I infrastructure, to address CSM,
- launching a ManuFuture trans-sectorial platform, to: develop SI and its implementation framework (FW) and manage it, to pursue CSM.

Since 2004 the ManuFuture platform has set up a rolling process to develop SI: i.e. the ManuFuture Vision 2020, the Strategic Research Agenda (SRA), the Roadmaps;

and the ManuFuture framework (FW) in which SI is implemented. It has managed the process, through basic activities and pilot actions and is contributing to the pursuit of CSM. Referring to the general reference model, (see figure 1.3) the ManuFuture platform may be considered as an *infrastructure*, acting within the innovation cycles. It manages the SI life cycle: i.e. from foresight, to roadmapping, implementation, monitoring of its implementation and effectiveness; and the ManuFuture framework life cycle, see fig. 6.3: i.e. from definition, to implementation, management, monitoring of its implementation and effectiveness, within the ManuFuture. All this is done to pursue CSM.

Models for developing SI and framework are reported in this chapter. Studies carried out within SRA, leading to Roadmaps, are reported in chapter 4. Chapter 5 presents the Roadmaps obtained referring to SRA models and the model described in paragraph 3.1.9. Chapter 6 reports to current ManuFuture framework derived from SRA studies. Finally, basic activities and pilot actions, within the ManuFuture framework are reported in chapter 6. The work done and the results obtained have been discussed and confirmed, in the last five ManuFuture Conferences [3.5 - 3.9] by stakeholders (industry, university, research institutes and centres, and financial institutions).

Within a ManuFuture framework being set and expanded, SI is being diffused, adopted and used by PAs in the EU, and at national and regional level, as well as the IK-Ts.

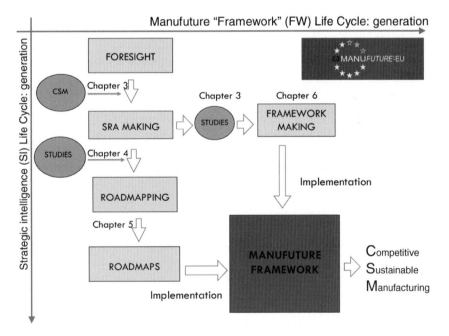

Fig. 3.2. Generation Phase of SI and FW: Their Interaction to Pursue CSM

These may be considered as part of the ManuFuture basic activities. Pilot initiatives, to focus and accelerate basic activities, are in progress. All the above is reported in chapter 6, together with perspectives.

Figure 3.2 shows the life cycle generation phase and interaction of both SI and ManuFuture frameworks and how this book is dealing with them.

The entire life cycle of SI, within the current ManuFuture framework, see figure 6.3, and the enabling and supporting basic activities and pilot actions are reported in chapter 6.

3.1.1 The ManuFuture Platform

The ManuFuture Technology Platform, as previously described, is the industry-led grouping of European initiatives and actions on European, national and regional levels, to assure the future of manufacturing in Europe.

Referring to the general reference model, see figure 1.3, the ManuFuture Platform may be considered as an infrastructure, acting within *the innovation cycles.*

As shown in figure 3.2, it manages: SI life cycle, i.e. from foresight, to roadmapping, implementation, monitoring; and framework life cycle: i.e. from definition, to implementation, management, monitoring; implementation of SI within ManuFuture framework (FW), to pursue CSM. ManuFuture Vision 2020, SRA and Roadmaps generation processes are reported hereafter [3.10].

3.1.2 The ManuFuture Vision 2020

A High Level Group of European executives, from research organisations and industry, invited by the European Commission, offered their expertise and insights as a basis for a structured debate, leading to a shared vision of the way ahead for EU manufacturing: ManuFuture Vision 2020 [3.11].

The key conclusions may be summarised as follows:

- There is a need for the development and implementation of a European Manufacturing strategy, based on research and innovation, following the ManuFuture approach.
- As each job in Manufacturing is linked to two jobs in services, the reliance on services cannot continue in the long-term without a competitive EU Manufacturing.
- A concerted effort will be needed to transform European Manufacturing from a resource intensive to a knowledge-intensive, innovative sector.
- The challenge is to move towards a new structure, which can be described as 'innovating production', founded on knowledge and capital.
- The knowledge-driven economy demands: a competitive and sustainable E&I&RTD system; a new approach to knowledge generation and innovation; adaptation of education and training schemes.

3.1.3 The ManuFuture Strategic Research Agenda (SRA)

The socio-economic and technological drivers, identified in the Manu-Future Vision document, are challenges for future European manufacturing. To address these challenges industry and policy-makers need to reconcile policies and approaches with the objectives of competitiveness and sustainable development. A time span of 10 to 15 years can bring about dramatic changes. It is not only industrial labour that is cheaper in some regions outside Europe, but also engineering and management.

Today, Europe still has the possibility and means to counteract this situation, but it must do so in a decisive, concentrated manner, based on sound strategic analysis, as for the Manu-Future Strategic Research Agenda [3.12]. It is based on three fundamentals: knowledge-based manufacturing, roadmap for industrial and E&RTD&I transformation multi-level action.

European Manufacturing has a huge potential for generating wealth, jobs and a better quality of life. It is generally knowledge-intensive, and embraces many different sectors. Still, as previously presented, a new industrial paradigm is needed.

This is HAV K-based Competitive Sustainable Manufacturing. The added value is the leverage to achieve dominance in markets, since purely cost-based competition, see figure 0.1, is not compatible with the goal of maintaining the community's social and sustainable values.

3.1.4 A Roadmap for Industrial Transformation

In the medium term, up to the 2020 time horizon of the ManuFuture vision, foresight studies indicate that to pursue HAV CSM, transformation of industry and the related E&RTD&I System should take place, being driven by the ESET context changes, as represented in RTD reference model shown in figure 3.3.

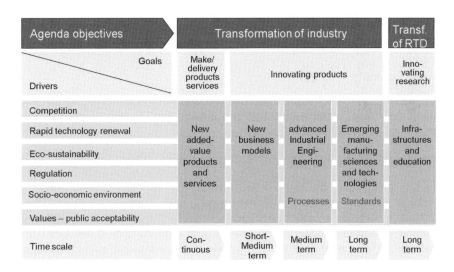

Fig. 3.3. Industry and E&RTD&I System Transformation: Reference Model [3.11]

The main drivers are:

- competition, especially from emerging economies,
- the shortening life cycle of enabling technologies,
- environmental and sustainability issues,

- socio-economic environments,
- regulatory climate,
- values and public acceptance.

The necessary E&RTD&I activities would be carried out, as shown by the RTD Reference Model, within five pillars of activity . They span the short to long-term time frame. The main features of the reference model are described hereafter.

Principal Drivers of Change

Competition

The context in which manufacturing companies work in the future will depend even more on flexibility and speed, as well as on localised production. Manufacturing is also likely to become increasingly service intensive. This service orientation of manufacturing and the increased customer demand will have consequences for the organisation of production, supply chain management and customer relations. Furthermore, there is a continuous increase in foreign direct investment in manufacturing outside Europe [3.10].

Rapid technology renewal

Especially in the fields of nanotechnology, materials science, electronics, mechatronics, ICT and biotechnology there will be rapid renewal. The development of new production processes based on research outcomes, and the integration of separate technologies exploiting the converging nature of scientific and technological developments, may radically change both the scope and scale of manufacturing [3.10].

Eco-sustainability

The manufacturing sector will also have to comply with stricter environmental regulations in the future, which should further stimulate the adoption of energy- and resource-saving technologies [3.10].

Socio-economic environment

Manufacturing in 2015 to 2020 will be called upon to provide solutions meeting new societal needs and the demands of an increasingly ageing public, having an impact on mobility, the size of the labour force, and on customer requirements. At the level of the labour supply, the manufacturing and research sectors will be confronted with the retirement of the currently large age groups, while innovation might require completely new sets of skills – the availability of which, in both manufacturing and research, could become a critical factor [3.10].

Regulations and standards

Stricter environmental and safety regulation will lead to changes in manufacturing. The intellectual property rights (IPR) system might have to respond to changes in an innovation process that is increasingly based on knowledge sharing and networking.

The adoption of new technologies in manufacturing will also depend on the availability of industrial standards and testing procedures [3.10].

Values – public acceptability

Recent debates on genetically modified food and stemcell research highlight the need to take ethical concerns into account when science and new technology are being adopted and exploited. At the same time, it should be noted that this could lead to Europe falling behind in some areas of technology [3.10].

3.1.5 Pillars and Domains of Activity

1. Transformation of industry

The transformation follows the RTD reference model and has to be realised by RTD in the fields of actions called pillars. (see figure 3.3)

1.A. New added-value products and services

European products are associated with high quality, appealing design and cutting-edge technologies. The effectiveness of the ManuFuture research agenda in transforming industry will depend upon manufacturers' readiness to leverage these strengths, while adapting continuously to change in an open, fast-moving global industrial environment.

1.B. Innovating production

A fundamental concept of the ManuFuture vision is that of innovating production which embraces new business models, new modes of industrial manufacturing engineering and the ability to profit from ground-breaking manufacturing sciences and technologies. Even the factories in which these new forms of production take place are regarded as complex, long-life products, operating with the latest technologies and adapting continuously to take account of customers' and market requirements. They may represent a European high-value product. A relevant example, see figure 3.4, has been developed within the EUROSHOE project [3.13]. It is a first example of customised shoes advanced factory, that at the same time, acts as an RTD&I laboratory. It is operated by SYNESIS (see par. 6.1.2). Customisation is the driving factor of the internal development of factories and causes changeable management, fast engineering and flexible and capable production of a high level of quality. Factories of the future are products – specialised on different areas – which operate permanently on the highest level of effectiveness (economic – ecological and social). Factories are the workplaces of millions of people in Europe and billions in the world. The culture of production in these factories is influenced by the regional environment. Europe's innovating production – the ManuFuture factory (figure 3.5) – includes the culture of management and innovation and makes it a trendsetter in the world.

Fig 3.4. Customised Shoes Advanced Factory Acting as a RTD&I Laboratory (SYNESIS)

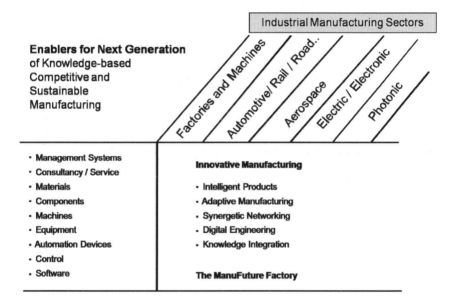

Fig. 3.5. The ManuFuture Factory as European High-Value Product [3.14]

2. Transformation of the E&RTD&I System

The initiative fosters also the transformation of RTD and education infrastructure for high-value manufacturing for a more and more efficient generation, distribution and use of knowledge in Europe and, specifically, in its regions.

Concentrations of such efforts will attract high-value manufacturing industry as well as the other fundamental actors such as universities and research centres even from outside Europe.

3.1.6 Multi-level Action

To attain the objectives set, it will be necessary to involve the largest possible number of stakeholders. Co-operation is critical between ManuFuture and the various existing and proposed technology platforms focusing on common goals and action plans – whether applied at EU or national /regional level and whether sectorial or technological in scope.

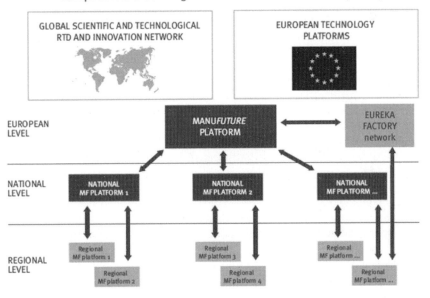

Fig. 3.6. Implementation of the SRA through Collective Action EMIRA [3.7]

European level

The ManuFuture Platform (figure 3.6), which represents a common asset for the whole European manufacturing and services industry, has to define, develop and consolidate innovative processes, methodologies and tools for all industrial sectors.

Those trans-sectorial deliverables will represent the common starting point for the activities with the other Technological Platforms (TP's), (figure 3.7), at national/regional level to address specific vertical needs.

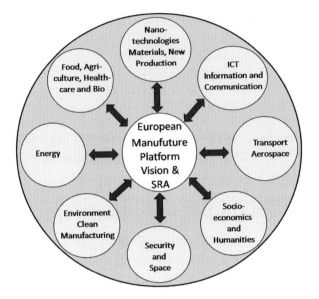

Fig. 3.7. ManuFuture and European Thematic Priorities [3.15]

National/regional level

National technology platforms related to the ManuFuture ETPs are being created in individual EU member states, all adopting the main development goals identified in both ManuFuture – a Vision for 2020 – and the current document.

Other initiatives can also stimulate the emergence at regional levels of equivalent concepts promoting competitiveness via synergy between sciences, education and industry. At the end, a coordinated effort at all levels is working with the aim of defining the manufacturing research priorities and committing to make it happen.

Aligning the development goals and priorities of all 27 member states is therefore crucial in building a common interest in close co-operation between production companies and RTD organisations as a foundation for expansion into global markets.

25 national ManuFuture initiatives have already been launched during 2007, through a spontaneous process. They are successful examples of efficient networking and collaboration, and a strong link with industry, research and education institutions and the regions in the European Union.

SME level

SME's are main players in several sectors, capable to develop, produce and sell innovative products and services to more and more demanding consumers. In others, they are linked in diverse networks with OEMs in the value chains.

3.1.7 Roadmaps: Generation Processes

Roadmaps, focusing on key enabling technologies, relate to strategic industry needs. Roadmapping was carried out by LEADERSHIP SSA [3.16] aiming at:

- enhancing the sectorial implementation of the ManuFuture SRA,
- incorporating all related SRAs research manufacturing contents, identified through exchanges with relevant European and ManuFuture national technology platforms co-ordinating at high level with existing and emerging roadmapping activities, also funded at European level, national and regional levels,
- developing implementation plans for a ManuFuture best practice and concept system,
- activating a consultation process through two annual targeted Roadmaps dissemination actions,
- timely delivering of Roadmaps and other actions,
- ensuring proper coordination of all the activities with strategic, steering and technical layers and management functions supported by web infrastructure,
- independent scientific advising.

The activities were carried out following the scheme shown in figure 3.8. The EU supported the development of Roadmaps for the implementation of the Strategic Research Agenda to find out the priorities of industries in a bottom-up process. Summarised results are presented in the chapters 4 and 5. All 25 sectors of manufacturers of machines, systems and consumer-near products agree in 90% of the proposed actions. The results have been discussed with ManuFuture related ETPs. There are some fields of action which are of common interest as well as sector specific.

Thus it was possible to find out trans-sectorial roads and sectorial roads with high communality. They deserve high priority and appropriate investments.

Leadership Framework

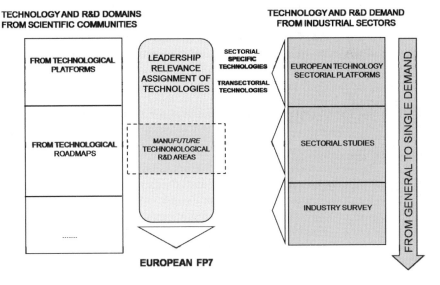

Fig. 3.8. Leadership Methodology [3.16]

3.2 Transformation of the E&RTD&I System: State of the Art and Prospects

To pursue HAV CSM, European industry – while acting in the globalising market - must undergo a transformation that should be enabled and supported by the E&RTD&I (the K-triangle) System. This, in turn, should become more and more robust and effective, competitive and sustainable, while getting global. Then, both industry and the E&RTD&I system –while interacting within the value chain linking science to innovation, to market – must undergo specific transformations, CSM- oriented.

Industry will transform, by developing new HAV products and services, processes and business models as well as adopting related enabling technologies. The E&RTD&I system transformation would concern both the new competences and enabling technologies that it would develop in order to work with and for industry and, on the other side, its structure. This would take place within EMIRA, a subset of ERA.

The European Research Area (ERA) is a political concept proposed by the commission and endorsed by the European parliament and council to overcome the present fragmentation of Europe's efforts in the area of research and innovation.

The concept comprises organising co-operation at different levels, co-ordinating national or European policies, networking teams and increasing the mobility of individuals and ideas. It is an area where the scientific capacity and material resources of the member states can be put to best use. An area that is open to the world where national and European policies can be implemented more coherently, and where people and knowledge can circulate more freely. The multiannual framework programmes are the financial instruments to implement the ERA.

Major changes in and outside Europe affecting ERA are: globalization of knowledge production, consensus on global challenges, enlargement of the EU [3.17]. Further, the emergence of new research locations and new research emphasis to maintain competitiveness, quality of life and assist developing nations has placed a greater emphasis on international scientific and technological co-operation. What is becoming more urgent is the need at all levels for co-ordination, coherence and visibility, including through leadership, in order to make Europe's international science and technology more effective, maintain the attractiveness of Europe as a place to do research[3.17]. In 2007 the European Commission published a green paper on ERA [3.18], reviewing progress made, where it still needs to be made and raising questions for debate. The public consultation has been very successful [3.19].

Consequently, the European Parliament and advisory bodies support the need for concerted action at European and national levels on six axes:

1. Realising a single labour market for researchers
2. Developing world-class research infrastructures
3. Strengthening research institutions
4. Sharing knowledge
5. Optimising research programmes and priorities
6. Opening to the world: international cooperation in S&T

The European economic and social committee highlights in particular the need for a European internal market for research and innovation. The committee of the regions stresses the important role of regions and cities for developing the European research area. The commission and member states are launching in 2008 new initiatives to develop ERA, including an enhanced political governance of ERA, called the "Ljubljana Process" [3.20], and five initiatives on specific areas of the ERA green paper.

ManuFuture SRA has dealt with E&RTD&I system transformation, within the aforementioned frame. In the following paragraphs, the state of the art of the main aspects of the E&RTD&I system is reviewed. Perspectives, according to ManuFuture, are reported.

3.2.1 Education

Education concerns competences necessary for the Research and Technology Development, Innovation, Market (RTD&I&M) value chain. Hence they encompass from researchers to industrial innovators, to manufacturing executives and also Ph.D's and students. The state of the art, concerning OECD countries and the European Union, as well as the ManuFuture SRA recommendations are summarised hereafter.

A - Trend in education at OECD level

The continued progression towards the knowledge-based economy and the expansion of knowledge-based industries has raised demand for professional workers of all kinds, including scientists and engineers. Across the OECD, growth of employment rates in professional occupations has outpaced employment growth overall, often by a wide margin [3.21].

Such growth demands highly educated workers with a variety of skills and can strain domestic education systems. Researchers are a particularly important set of professionals because they serve as a central element of RTD systems and require specialised training, often extending over many years (tertiary level and beyond).

In 2002, approximately 3.6 million researchers were engaged in RTD in the OECD area, up from 2.3 million in 1990 [3.21]. This corresponds to about 8.3 researchers per 1000 employees, a significant increase from the 1995 level of seven researchers per 1000 employees. Out of these 3.6 million researchers, most were engaged in the business sector and just over 25% were engaged in the higher education sector [3.21].

A.1. Ph.D.

In 2002, OECD universities awarded some 5.9 million degrees at university level, among the 156 000 doctorates. In other words: fewer than one person of three at the typical age of graduation completed a university degree, while one out of 100 received a doctoral degree (figure 3.9) [3.21]. Finland and Australia had the highest graduation rates at university level (over 45% of the population), and Sweden and Switzerland the highest rates at doctorate level with 2.8 and 2.5 doctorates per 100 at the age of graduation, respectively. The gap widens for doctoral degrees: European universities awarded 55% of all S&E doctorates. [3.21]

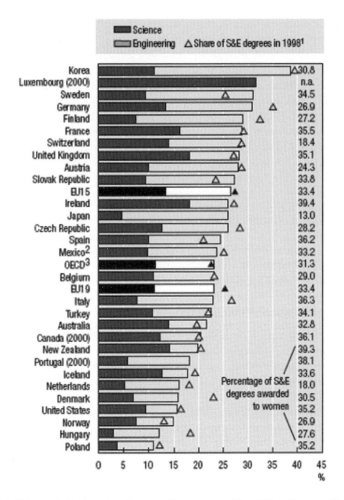

Fig. 3.9. Science & Engineering Degrees, Percentage of Total New Degrees [3.21]

A.2. Graduates and Students

Graduates

Since 1998, the numbers of science and engineering graduates have continued to increase. Overall, about 23% of the 5.9 million degrees granted at universities in the OECD area were granted in science and engineering. There are, however, important differences among countries in terms of starting points as well as the evolution of the supply of science and engineering graduates. This generally reflects the industrial structure, historical academic traditions, but also higher education and research funding policies [3.21, 3.22].

The supply of science and engineering graduates continues to expand in absolute terms, but in the EU between 1998 and 2004, Denmark, Italy, Germany, Hungary and

Finland experienced a drop in the share of university graduates with science and engineering degrees, as did Korea and the United States. Further exacerbating the situation in the United States is a decline in first time, full-time enrolments of foreign PhD students, which fell for the second consecutive year in 2003.

Irrespective of their own recent declines, EU countries still produce a greater share of science and engineering graduates than Japan or the United States, despite the smaller share of researchers in the workforce: 27% of EU university graduates (see figure 3.10) obtain a science or engineering degree compared to 24% in Japan and just 16% in the United States [3.21]. Since 2004, China has produced three times more graduates in engineering than the US, and India produces almost the same number of engineers as the US. China has close to the same number of full time researchers as all EU Member States together and India has the largest pool of young university graduates in the world [3.23].

Students

New university graduates are an indicator of a country's potential for assimilating, developing and diffusing advanced knowledge and supplying the labour market with highly skilled workers.

Scientific studies (excluding health and welfare) remain the second most popular field although the share of OECD graduates obtaining a science and engineering degree has declined to one fifth. In Korea science and engineering degrees account for almost 40% of all new degrees.

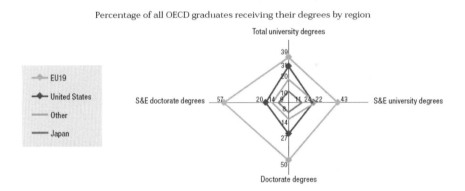

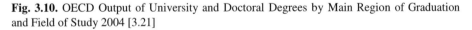

Fig. 3.10. OECD Output of University and Doctoral Degrees by Main Region of Graduation and Field of Study 2004 [3.21]

New graduates in science and engineering are 13% of population aged 20-29 in Europe and in Japan (2004 data), but only 10% in US (2004 data) [3.23].

Europe is still the major provider of science and engineering university degrees (43% of the total share in 2004) followed by emerging countries and United States (see figure 3.10) however, at the same time a decline of the share of science and engineering graduates in the EU and the ageing of a significant portion of the science and engineering workforce is a growing concern in many member states [3.21, 3.17].

In about two thirds of OECD countries, universities deliver more engineering than science degrees; in Finland, Japan, Korea and Sweden the number of engineering degrees awarded far exceeds that of science degrees.

B - Trend in education at EU level

Education and training have a critical impact on economic and social outcomes. Investment in education amounts to 5.5% of GDP or 500 BEURO each year [3.24].

Europe's tertiary education institutions produced close on 2.5 million new graduates in 2003 in the EU. This compared with just over 1 million new graduates in Japan and over 2.3 million in the United States. Comparing these new graduates against the young population, around 48 new graduates come out of one thousand people aged 20-29 in the EU. The stock of human resources in science and technology (HRST) is growing over time [3.24].

Table 3.1. Education in Manufacturing [3.21]

Education in Manufacturing	Total in engineering, manufacturing and construction	% of all students	% of all doctorate students	% of all graduates	% of all doctorate graduates	% of all PhD students
Students participating in tertiary education, aged 20 – 29 -2003 – EU 25	2036957s	14.3%				
Doctorate students (ISCED level 6), aged 20-29 –2003– EU 25	57834 s		16.2%			
Graduates from tertiary education aged 20-29 –2003– EU 25	313750 s			12.8%		
Doctorate graduates (ISCED level 6) aged 25-29 and selected countries –2003– EU 25	9433 s				16.0%	
PhD students, -2004– EU 25	65737					16.4%

B.1. Ph.D. at EU level

The EU dominates in the 'production' of PhD graduates overall as well as in the supply of PhDs in science and engineering. Out of a total of 156 190 PhD graduates in the OECD area in 2002, the EU19 (EU15 plus Poland, Czech Republic, Hungary and Slovak Republic) accounted for 51% while the United States represented 28% and Japan 13% (figure 3.11). In science and engineering, the EU also 'produces' more PhDs (55%) than the United States (25%) or Japan (9%) [3.22].

	PhD students		Science, mathematics and computing		Engineering, manufacturing and construction	
	(number)	Female (% of total)	(number)	(% of all PhD students)	(number)	(% of all PhD students)
EU-25	401 386	46.6	85 547	21.3	65 737	16.4
EU-15	322 924	47.4	71 168	22.0	48 232	14.9
Euro area	205 993	49.0	41 772	20.3	29 346	14.2

Fig. 3.11. PhD students in Europe - 2004

B.2. Graduates and students at EU level

In 2003, over 14 million persons in EU-25 were in tertiary education, and more than 20% of them were aged between 20 and 29.

The overall number of tertiary level students is growing. This tendency is true for men and women. Between 1998 and 2003, the number of people in tertiary education in EU-25 grew at an annual average rate of 5% for male students and of up to 6% for female students.

Meanwhile, the new member states had the highest growth rates in comparison to the other EU countries.

One student out of four was following courses either in "science, mathematics and computing" or in "engineering, manufacturing and construction" in the EU-25 in 2003. Nevertheless, engineering courses were marginally more popular (14.3%) than science (10.6 %). Ireland had one of the highest proportions of students studying science (14.1%), while the highest proportion was in Germany (14.6%). For engineering courses, Finland had the highest proportion of students (26.6%) [3.22].

C - ManuFuture recommendations [3.12]

In order to create the strategic European infrastructure for competitiveness – the foreseen *Knowledge Industry Fabric* – it is essential to prepare people trained and educated to serve the scope of new manufacturing, i.e. the core condition to generate industrial HAV.

People are the strategic asset that the European research system should prepare to continuously apply and transform knowledge into a competitive tool producing HAV for European manufacturing.

This objective meets the political European goal which declares the knowledge triangle of research, education and innovation to function within favourable framework conditions which reward the knowledge that input to work.

Knowledge-based production requires the support of new kinds of education and training schemes integrating research with technology and manufacturing.
The priorities are to:

* reorganise educational programmes,
* establish Europe-wide educational systems,
* address the workforce as a societal issue.

Integrating education and industry is mandatory. A highly promising approach would be to integrate the factory environment with the classroom, as being proposed and implemented by the 'Teaching Factory' and the 'Learning Factory'.

New forms of basic and life-long training, moving beyond the traditional disciplinary boundaries, should be envisaged. Human-oriented machine interfaces can support workers' and users' skill development for handling the complex technical systems of intelligent manufacturing. New education and training goals in the global economy are:

- define and comprehend the needs of the manufacturing industry for training and education in the years to come,
- create a European framework for pilot implementation of the 'Teaching Factory' and the "Learning Factory",
- develop the ManuFuture educational system, with engineering curricula embedding entrepreneurship and innovative spirit, to harmonise the manufacturing qualifications of EU member states,
- retrieve the respect and appeal of manufacturing, in order to attract young people as the workforce to give society the capacity to achieve sustainability in manufacturing.

D - Pilot initiatives on education

Some preliminary best practices, complying with ManuFuture recommendations, are outlined hereafter (see also paragraph 6.2.1). They should contribute to improve the education-RTD-innovation-market value chain.

Learning Factory. The objective is to promote knowledge, competences and best practices for advanced industry, by integrating learning, research, innovation activities, through a context-aware virtual factory for collaborative learning. This is needed (figure 6.6), to promote and support a HAV knowledge-based CSM industry.

Teaching Factory. The objective is to seamlessly integrate education, research, innovation activities within a single initiative, to develop competences and skills needed to promote and support future perspectives of a HAV knowledge-based CSM industry (see figure 6.7).

3.2.2 RTD&I System

RTD&I concerns the knowledge generation process, necessary to the Research and Technological Development, Innovation and Market (RTD&I&M) value chain. The state of the art, concerning OECD countries and the European Union, and the ManuFuture SRA recommendations may be summarised as follows.

A. RTD at OECD level

At OECD level the manufacturing sector remains an important locus of innovation and continues to account for the largest part of RTD spending in most OECD countries. Within manufacturing, the high- and medium-high-technology sectors continue to grow in terms of their contribution to Added-Value and total RTD performance, related to medium-low- and low technology industries (figure 3.12).

In the OECD area, high-tech industries (including aerospace, pharmaceuticals, computer equipment, radio, TV and communication equipment, as well as instruments) account for more than 53% of total manufacturing RTD, but regional differences remain.

In 2002, RTD in high-tech industries accounted for over 60% of total manufacturing RTD in the United States compared to 48% and 46% in the European Union and Japan, respectively.

Manufacturing RTD expenditure is primarily oriented towards high-technology industries in Ireland, Canada and Finland. Medium-high technology industries, such as machinery and chemicals, account for 50% of RTD or more in the Czech Republic and Germany. Norway is the only OECD country in which medium-low and low-technology industries account for more than 40% of manufacturing RTD.

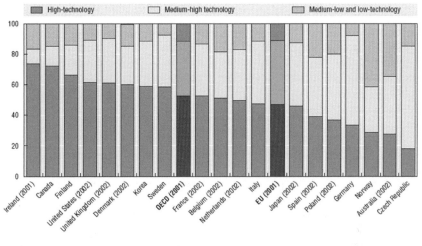

Source: OECD STAN Database 2005.

Fig. 3.12. Share of Business RTD in the Manufacturing Sector by Technological Intensity, 2003 [3.21]

Most emerging countries that have successfully attracted foreign RTD, motivated by the success of developed countries, have put in place policies to strengthen their national innovation system.

Innovative RTD and technology sourcing are indeed still undertaken predominantly in developed countries, mainly because of the presence of world-class clusters of technological and industrial activity including centres of excellence and an effective national innovation system.

On the other side, there is increasing attractiveness for foreign RTD of some emerging countries, for several reasons, including proximity to new manufacturing activities.

A.1. Researchers at OECD level

In the vast majority of OECD countries, the number of researchers rises at a faster rate than the number of total R&D personnel. This is partly due to the increased number of postgraduate students who perform R&D and are counted as researchers in the higher education sector.

Greater use of new information technologies in R&D activities may also explain the need for fewer technicians and support staff per full-time equivalent researcher. The number of researchers has increased the most in China (from a small base), Finland and New Zealand, with average annual growth rates of close to 9%, more than double the OECD average of 3.2%. Business enterprise researchers continue to account for the bulk of the researcher population.

In 2002, some 64% of all researchers in OECD countries (or 2.2 million of the total 3.4 million) worked in the business sector, a figure that remained fairly constant over the previous decade. Nevertheless, there are clear regional differences (figure 3.13).

Researchers to labour force ratio is greater in Japan (10.4 researchers per 1000 labour force) and the United States (9.6 per 1000) than in the European Union-25 (5.8 per 1 000). Many countries expect the ratio to increase. The US bureau of labour statistics estimates that scientific and engineering (S&E) occupations will increase by 26% in 2012 compared to 15% for all other occupations [3.21, 3.22].

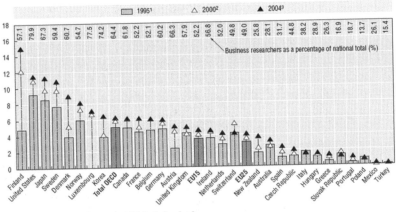

1. 1993 instead of 1995 for Austria; 1996 for Switzerland.
2. 1999 instead of 2000 for Sweden, Denmark, Norway, New Zealand and Mexico; 1998 for Austria.
3. 2002 for Austria, Canada, Turkey, United States and OECD total; 2003 for Australia, France, Germany, Greece, Italy, Mexico, New Zealand, Norway, Sweden, EU25 and EU15.

Fig. 3.13. Business Researchers per Thousand Employment in Industry [3.21]

A.2. Employed at OECD level

Human resources in science and technology (HRST) are the main pillar of knowledge-based economies. People working in HRST occupations represent 25% to 35% of total employment in OECD countries. Demand for HRST has never been higher: employment in HRST occupations grew twice as fast as overall employment between 1995 and 2004 in most OECD countries.

The number of researchers in OECD countries, a subset of HRST, grew from 5.8 researchers per 1 000 employees in 1995 to 6.9 per 1000 in 2002 [3.21].

Over the past decade, employment in HRST occupations has grown much faster than total employment in all countries, at an average annual rate of 2.5% in the United States, 3.3% in the EU15, 4.1% in Korea and 4.5% in Australia. Some countries with low shares of professionals and technicians have been catching up (e.g. Spain, Hungary, Ireland and Greece) [3.21].

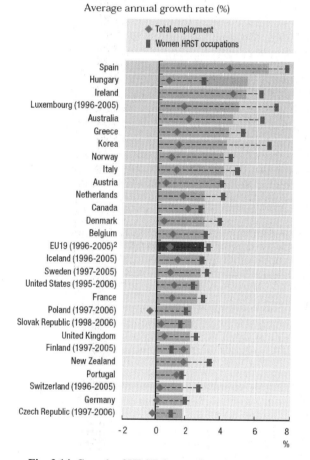

Fig. 3.14. Growth of HRST Occupation 1996-2006 [3.21]

Since the 80s, growth in science and engineering occupations in the United States has been more than four times the growth rate of all other occupations (NSF, 2006) (Figure 3.14). In Europe, efforts to increase spending on RTD and innovation will further increase demand for HRST. According to EU estimates, meeting the EU Lisbon/Barcelona targets of 3% of GDP for RTD will require another 700 000 researchers [3.21, 3.22].

B. RTD at EU level

At European level according to the Eurostat Yearbook 2006, 981.209 EU-15 re-searchers (RSE) are professionals employed by this system. This figure includes managers and administrators engaged in the planning and management of the scientific and technical aspects of a researcher's work as well as postgraduate students engaged in RTD. Considering the amount of enterprises and workers operating in manufacturing sectors, the percentage researchers/manufacturing workers is around 3.6 [3.24].

Manufacturing is for EU-25 by far the most important sector of activity for RTD expenditure and personnel accounting for 82%, followed by services with about 17%. With EUR 34.6 billion, Germany was ahead in absolute terms in manufacturing, whereas the United Kingdom ranked second (EUR 15.2 billion).

Two thirds of all European R&D is conducted by the private sector, and is concentrated in a very limited number of large companies. However, Europe still lags behind the US and Japan in terms of business expenditure on research and development [3.17].

B.1. Researchers at EU level

The number of researchers in full time equivalent (FTE) per thousand labour force amounted to 5.4 in the EU in 2003, compared to 10 and 9 in Japan and the US respectively and remains essentially unchanged since 1999 [3.17].

Nonetheless, the number of researchers per 1000 workforce in the EU has been growing at an average annual rate of 2.8% between 1997 and 2003. Only few member states showed a negative or slow growth rate. Data for 2004 show that the share of researchers in the workforce is slightly up (+ 3.5%) compared to the past average [3.17]. Total employment in EU-25 was in R&D in 2004, the head count being 2.82 million people.

The importance of high-technology sectors has increased considerably over the last few years and this has had a significant impact on the structure and organisation of employment in Europe. In 2004, 53.4% of R&D personnel in EU-25 were employed in the business enterprise sector, 31.1% in the higher education sector and 14.3% in the government sector.

In the EU-25, 572 951 researchers measured in FTE were employed in the business enterprise sector in 2003. The largest share of these business researchers worked in the manufacturing sectors (413 340 persons).

This trend can also be observed in the individual EU-25 countries. In absolute terms, Germany had the highest number of BES researchers in Europe, with 161 980 persons, followed by the United Kingdom, with 102 684 persons. The proportion of these researchers working in the manufacturing sector reached 88% in Germany and 76% in Sweden.

The deficit in the share of researchers of the workforce as compared to the US and Japan is mainly located in the business sector. Of the estimated total of 1,180,000 researchers (FTE) in the EU-25 in 2003, about 50% were employed in the business sector. This compared to some 68% in Japan and about 80% in the US [3.17]. EU countries still train more researchers for a science and engineering doctorate than the US and Japan.

The EU shows some serious levels of unemployment among researchers, and the lower salary levels of researchers in comparison with other employment sectors, this would seem to indicate that there is no shortage of researchers either [3.17].

Imbalances between national labour markets also cause a drain of researchers to other countries, including outside the EU, in particular to the US. Although an estimated 80,000 to 100,000 EU born researchers (in head count) are active in research in the US.

This only amounts to some 5% to 8% of the total EU researchers' population. Set against the concept of a beneficial 'brain circulation', such a contingent of internationally mobile researchers would even be desirable if there was a clear prospect that a large portion of this group would (eventually) return to the EU [3.17].

B.2. Employed at EU level: Skills and employment in manufacturing at European level

In 2004, more than 34 million were employed in manufacturing in EU 25. Of the 130 million jobs in services in EU-25, half of these were in know-ledge-intensive services (KIS) and the other half in less knowledge-intensive services (LKIS). [3.24] From the 34 million people that were employed in manufacturing, 11 million were in medium-high-tech manufacturing (5.7% of total employment) and more than 2.2 million in high-tech manufacturing (1.1% of total employment).

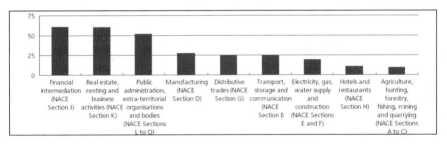

Fig. 3.15. Human Resources Working in Science and Technology Occupations, Breakdown by Activity, EU-25, 2004 (% of Sectoral Employment)

Of the total workforce in manufacturing and services of 166 million, almost 20 million persons were employed in high-tech manufacturing and services within the EU in 2004. [3.24]

In Europe, almost two thirds of HRST were concentrated in the four largest economies (22% in Germany, 12% in both France and the United Kingdom, 11% in Italy). Northern European countries were among the top ten with respect to the share of science and technology occupations in total employment (more than 35%); in Spain, Greece, Ireland and Portugal the share was around 20%.

In general, the supply in Human Resources in Science and Technology (HRST) has increased as inflows of graduates from tertiary education have grown. Germany, the UK and France had the highest number of HRST in 2004, with more than 10 million HRST in each country. These three EU countries together accounted for nearly half of the EU's 76 million total HRST. The number of persons employed in science and technology (HRSTO) increased between 1999 and 2004.

Table 3.2. Human Resources in Science and Technology Intensity (HRSTE) of Employed People With S&T Education – as a Percentage of Total Employment, 25-64 Years Old – 2004 [3.22]

	High-Tech Manufacturing	Medium High-Tech Manufacturing	Medium Low-Tech Manufacturing	Low-Tech Manufacturing
(HRSTE) - EU 25	32.6 %	24.0 %	13.7 %	13.5 %
(HRSTE) - EU 15	35.1 %	25.5 %	14.5 %	14.7 %

The breakdown of employment in science and technology in 2004 Sweden had one of the highest proportions of its working population in S&T occupations having completed tertiary level education in 2004 (around 21%). The 498.000 persons employed in science and technology with tertiary level education reached a proportion of less than 10 % of the total labour force. If the people working in science and technology without tertiary level education are included, this proportion only goes up to 15%. The highest representation of scientists and engineers in 2004 is found in Belgium, where 7.5% of the labour force declared that they had an occupation qualifying them as science engineers.

Table 3.3. Human Resources in Science and Technology (HRST), EU-25, 2005 (Thousands) in Manufacturing

(HRST) – TOTAL	10 096
1 – HRST CORE	3 353
2 – HRST – EDUCATION	6 163
3 – HRST – OCCUPATION	7 286
4 – SCIENTISTS AND ENGINEERS	1783

(1)Those people who have successfully completed the third level of a science and technology field of study and who are employed in a science and technology occupation.
(2) Those people who have successfully completed the third level of a science and technology field of study.
(3) Those people who are employed in a science and technology occupation.
(4) Those people who work in physical, mathematical and engineering occupations or in life science and health occupations.

Not surprisingly, high-technology manufacturing which includes manufacture of office machinery, computers, radio, television, medical, precision and optical instruments, watches and clocks was the most knowledge-intensive of the manufacturing industries in the EU in 2004, where around one third of all employed people had tertiary science and technology education.

As expected, medium-low technology manufacturing and low technology manufacturing scored lowest, with EU rates reaching only 13.7% and 13.5% respectively.

3.2.3 E&RTD&I Investments, Programmes and Initiatives Towards CSM

The European research area covers from European to national and regional level. The European "army" of HRSTO; the high number of: education, research and technological development, innovation organizations, making complex but effective networks (see figure 3.16); quite high investments in E&RTD&I; ongoing and perspective related programmes and initiatives; make ERA and EMIRA one of the best E&RTD&I system in the world. Investment programmes and initiatives will be dealt hereafter.

European Collaborative Research Networks

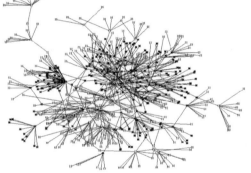

Fig. 3.16. European Collaborative Research Networks within ERA [3.15]

A. European public and private investments in RTD manufacturing

In Europe, industry carries out close to two thirds of all R&D activities. According to official R&D statistics for the EU-27, expenditure on R&D is distributed as follows: business (BERD) represented 128 billion in 2005 (64% of total), universities and higher education research and development spending (HERD) 44 billion (22%) and governmental research organisations spending (GOVERD) 26 billion (14%) [3.17].

At the EU-15 level, industrial production and technology (6.6%) had the highest rates of budget increase between 1999 and 2004 for all European socio-economic objectives. Belgium (33.3%), Ireland (27.1%), Finland (25.9%) and Spain (23.4%) allocated a large part of their total government RTD budget to this objective. In 2004, the objective 'industrial production and technology' followed and accounted for

11.4% of total European government budget appropriations or outlays for research and development (GBAORD). By taking into account that in 2004 the European Union allocated almost EUR 78 billion to GBAORD this implies an investment of 8.892 MEURO for industrial production and technology in EU-25 area.

For the EU-25 member states for which data are available, more than 90% of total business RTD expenditure was spent in high and medium-high-tech manufacturing in Germany, Hungary and in the United Kingdom. In general, the proportion of researchers among RTD personnel was higher in high-tech manufacturing than in total manufacturing. Hungary had the highest proportion with 85.8% of researchers in high-tech manufacturing.

Table 3.4. R&D Governmental Budget for Manufacturing Research [3.21]

	EU-15 (2004)	US (2004)	JP (2004)
INDUSTRIAL PRODUCTION AND TECHNOLOGYAPPROPRIATION AS % GBAORD	11,4	0,4	7,1

ANNUAL AVERAGE REAL GROWTH RATE (AAGR) OF GBAORD	1994-1999	1999-2004
INDUSTRIAL PRODUCTION AND TECHNOLOGY EU-15	-1.7	6.06
INDUSTRIAL PRODUCTION AND TECHNOLOGY JAPAN	19.05	5.08
INDUSTRIAL PRODUCTION AND TECHNOLOGY UNITED STATES	0.09	1.03

Europe has more than its share of the world's top 50 R&D investors: 20 are European companies, from which six rank among the ten fastest R&D growers.120 large European companies' performance levels of R&D investment are comparable to those of their counterparts outside the EU. The total BERD in Europe – which is not only determined by the large enterprise groups remains, however, relatively low in comparison to that of the US or Japan [3.17].

B. EU RFP and Eureka initiative

At European level on the basis of the Treaty of Amsterdam [3.25], all measures taken in the field of research funding and technological development are pooled in a community Research Framework Programme. In addition to the EU's Research Framework Programme, which is administered centrally by EC and under which calls for proposals are issued for topics determined in a top-down approach, other programmes (i.e. EUREKA) serve to promote and support European research co-operation.

B.1. EU Research Framework Programmes (RFPs)

The EU Research Framework Programme (RFP) is the largest funding programme for research projects in the world. The primary goal of the EU Research Framework Programme is to strengthen the scientific and technological basis for the community's industry and to foster its international competitiveness as well as to support all research efforts which are considered necessary for other policies of the community. This would support and foster the European Research Area (ERA). Objectives and actions vary from one FP to another.

The Seventh Framework Programme for research and technological development (FP7) is the European Union's chief instrument for funding research over the period 2007 to 2013.The total budget for FP7 is 51 BEURO over seven years. It bundles all research-related EU initiatives together under a common roof playing a crucial role in reaching the goals of growth, competitiveness and employment. As was the case for FP6, its main objective is to further the construction of the European Research Area. Its specific goals are:

- to gain leadership in key scientific and technology areas,
- to stimulate the creativity and excellence of European research,
- to develop and strengthen the human potential of European research,
- to enhance research and innovation capacity throughout Europe.

Further several schemes and instruments [3.26] have been launched and are being applied to foster co-operation and co-ordination of research activities. They are reported hereafter.

ERA NET Scheme. Its objective is to step up the co-operation and co-ordination of research activities carried out at national or regional level in the member states and associated states through: the networking of research activities conducted at national or regional level, and the mutual launching of national and regional research programmes. The scheme will contribute to making a reality of the European Research Area by improving the coherence and co-ordination across Europe of such research programmes. The scheme will also enable national systems to take on tasks collectively that they would not have been able to tackle inde-pendently.

Article 169. It enables the community to participate in research programmes undertaken jointly by several member states, including participation in the structures created for the execution of national programmes. Article 169 is potentially a very powerful instrument: Integrated Projects and Networks of Excellence integrate individual performers of research; Article 169 integrates national programmes. Article 169 is adopted by a co-decision process between the European Parliament and the council. The originality of Article 169 is related to the fact that the proposal comes from the Member States.

Actions according to Art. 171 EC JTIs. They are Joint Technology Initiatives (JTIs) and a new financial instrument of FP 7 adopted by the council based on Article 171 EC Treaty. Following Art. 171 the community may set up joint undertakings or any other structure necessary for the efficient execution of community research, technological develop-ment and demonstration programmes. Joint Technology Initiatives are private-public enterprises on a long term basis providing a combination of private and public funding for selected fields of technology. The aim is to increase the competitiveness of the European industry in these technology fields. The initiatives mainly result from the European technology platforms set up during the duration of FP 6. Up to now the council adopted five JTIs which can now start their work:

- Innovative Medicines Initiative (IMI),
- Nanoelectronics (European Nanoelectronics Initiative Advisory Council - ENIAC),
- Embedded systems (Advanced Research and Technology for Embedded Intelligence Systems - ARTEMIS),
- Air Transport(Clean Sky),
- Fuel Cells and Hydrogen (FCH),

The community's financial contribution is provided by the budget of the specific programme 'co-operation'.

B.2. Eureka initiative

Further, EU FPs and the EUREKA Initiative are the most relevant infrastructures promoting and supporting at European level– through RTD&I - the competitiveness and sustainability of the Science, Technology, Industrial Innovation, Market value chain.

Created as an intergovernmental initiative in 1985, EUREKA [3.28] aims to enhance European competitiveness through its support to businesses, research centres and universities who carry out pan-European projects to develop innovative products, processes and services. Thus the initiative is a complementary activity to programmes and measures initiated at the European Union level in this area. Up until now, EUREKA has supported more than 2,300 industrial research projects. EUREKA projects have attracted investment of over 23 BEURO of public and private money.

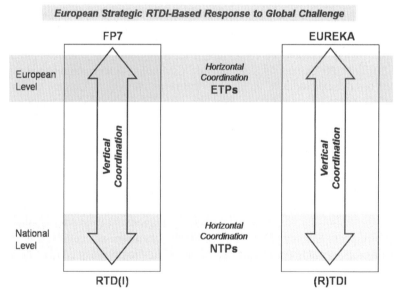

Fig. 3.17. FP7 and Eureka Initiatives: European Strategic Response to Global Challenge [3.27]

More than 12,000 partners have been involved in EUREKA projects, more than 42% of participants are SMEs. Through its flexible and decentralised network EUREKA offers partners rapid access to a wealth of knowledge, skills and expertise across Europe and facilitates access to national public and private funding schemes. EUREKA operates [3.28] through individual projects and strategic initiatives: i.e. Clusters, Umbrellas, Eurostars, see Figure 3.18.

Each year hundreds of individual projects are initiated by European companies, an increasing number of which are SMEs. These contribute to improved wellbeing, security, environment and employment in Europe and beyond.

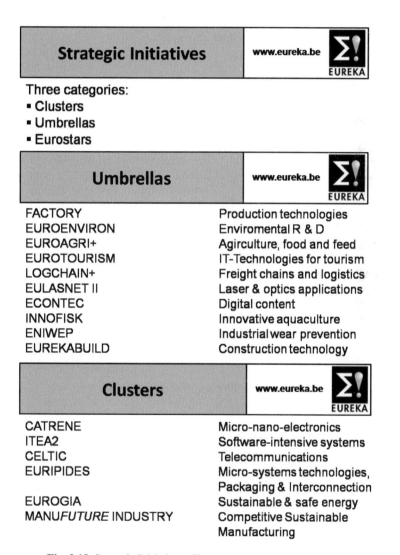

Fig. 3.18. Strategic Initiatives, Clusters and Umbrellas [3.29]

Clusters play a key role in building European competitiveness, driving European standards and the interoperability of products in a wide range of sectors. The result is a clear demonstration of the strength of pan-European teamwork in the European Research Area. Through its clusters, initiated and driven by industry, EUREKA offers a strong basis for trans-boundary R&D co-operation between large companies, SMEs and research institutes, with the direct involvement of national authorities.

Umbrellas are thematic networks which focus on a specific technology area or business sector. The main goal of an Umbrella is to facilitate the generation of EUREKA projects in its own target area. EUREKA's umbrella actions have proved to be effective as frontline networks and provide the proactive mechanism to generate SME projects. Umbrellas present a real opportunity to facilitate the development of high quality EUREKA and Eurostars [3.30] projects.

It will achieve the long sought objective of synchronised and secured national funding, adding commission co-funding and the strength of a central, single evaluation of proposals.

The Eurostars Programme allows SMEs to start application- oriented, transnational R&D projects and to exploit the results on the market on a short-term time scale, something that is of utmost importance to them.

The new EUREKA strategic document [3.31] fosters closer cooperation with the EU, national and regional initiatives like the Joint Technology Initiatives and European Technology Platforms. EUREKA should act as a facilitator for new activities like the implementation of Strategic Research Agendas (SRA) of European Technology Platforms (ETP) – see Manufuture Industry EUREKA Cluster, par 6.2.2 – Joint Technology Initiatives (JTIs) and other forms of international R&D cooperation, including that of regional clusters.

EUREKA is making an important contribution to the ERA through its experience in fostering large strategic projects (i.e. clusters) of high value. Their experience has positively contributed and will contribute to the development of current and future Joint Technology Initiatives.

C. ERA action for manufacturing

For European industry to move towards CSM, in the increasingly complex global ESET context, it is crucial that it modernises its manufacturing base and strengthens the links with academia, research and innovation.

This involves clear recognitions of the mutual value in an intimate collaboration and would rely on an effective and globally competitive European RTD&I internal market, such as that foreseen and fostered by the EC: the European Research Area (ERA), that in our case is EMIRA: i.e. the European Manufacturing Innovation and Research Area, that includes Education. Concentrating the forces for the structural change makes it necessary to join the stakeholders in a network of competences and excellences on European and national level. Three groups of stakeholders are involved in the innovation process (figure 1.16) and form centres of competences by overcoming the gaps between education, research and industrial enterprises. This is called the Knowledge Triangle.

Research done by industries in their research centres, in the institutes at the universities and public/private organisations and the education system represent a triangle for the innovation system. The innovation process supported by EMIRA is oriented to fasten the speed of innovation by a concept of high efficiency and synergy. The fact of high diversity and divergent regional cultures makes it necessary to define a new European way for highest skill, High-Adding-Value (industries) and efficient but application-oriented research in regional clusters.

EMIRA is the new way of innovation in manufacturing. Regional clusters follow the ManuFuture strategy and are connected to regional industries with their specific structures. Objectives of regional clusters are the creation of knowledge, continuous innovation and transfer of knowledge-based manufacturing in praxis by application oriented research, training and education in medium and high levels.

The accelerating globalisation of RTD&I and the emergence of new scientific as well as technological powers – notably China and India – are challenging the European Research Area, its competitiveness and sustainability.

ERA and its stakeholders – from infrastructures: such as RTD&I programmes and initiatives, Technology Platforms (TPs), at various levels (EU, member states, regions); to actors, such as universities, research institutes and centres, companies – represent a very powerful RTD&I system.

They must react swiftly and effectively to prevent being undermined by the incoming global technological and industrial revolution.

D. ManuFuture priorities for RTD&I

While the above process is taking place, ManuFuture lists a number of priorities for changes in the form and management of RTD&I infrastructures [3.12]. They should:

- foster an entrepreneurial culture to deliver and develop the marketing and exploitation of research results and schemes to help the creation of knowledge-based SMEs,
- create research infrastructures, mainly through networking, with regard to manufacturing research needs and the already existing infrastructures,
- establish favourable framework conditions to create an attractive fiscal environment at EU level,
- promote involvement of SMEs with RTD centres by co-financing multidisciplinary research programmes and knowledge transfer mechanisms.

This transformation requires new and dynamic research and innovation networks that must be nurtured to stimulate knowledge generation and ensure efficient transfer of its benefits to the manufacturing sector [3.26]. This involves clear recognitions of the mutual value in an intimate collaboration between the academic and industrial communities and knowledge transfer intermediaries.

3.3 Investment in RTD for Manufacturing: Boosting Economic Potentials

European Manufacturing with its industrial sectors creates an added value of 1.630 BEURO each year [1.14]. It is the backbone of the European economy. Investment in RTD is a strong contribution to CSD by opening the potential for future technologies. People and enterprises pay taxes to the governments. They expect a responsibility for the sustainability of the economy and welfare. Governments can push the speed of development by funding research and education for manufacturing. They even contribute to economic, social and environmental regulations for the future of industries. Regulations should have a scientific base.

RTD is the key issue for CSD and the development of future technologies and the transformation of the industrial system. In total, about 80% of the expenditures for RTD are spent by enterprises and 20% by governmental and other funding organisations. The governments play a key role, because of their non profitability and non commercial orientation. They can influence the common know-ledge base as well as long term developments.

Governmental support of RTD (figure 3.19) is necessary for:

- basic research for innovation,
- pushing industries to a higher competition by precompetitive application oriented research,
- scientific base of regulations and policies,
- analysis of the consequences of technologies and their impacts,
- acceleration of the diffusion of technologies by education and transfer,
- support of structural changes.

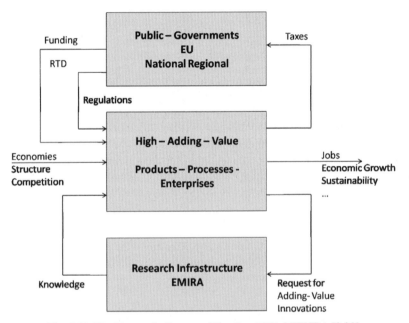

Fig. 3.19. The Economic System of Funding RTD ©IFF/IPA [3.32]

The main question for industrial expenditures is how profitable the investment of public and private funding for RTD is. Industry pays ca. 80% of RTD expenses and 20% is funded by public organisations and governments. About 60% of enterprises spend lower than 1% of their turnover for RTD in manufacturing.

Many companies especially in sectors of mass production are vulnerable and threatened by migration processes of low-cost regions. Others are even able to spend more but prefer short time profitability. European industries need a climate of innovation (figure 3.20) pushed by research and financial perspectives.

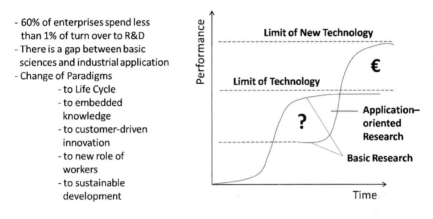

- 60% of enterprises spend less than 1% of turn over to R&D
- There is a gap between basic sciences and industrial application
- Change of Paradigms
  - to Life Cycle
  - to embedded knowledge
  - to customer-driven innovation
  - to new role of workers
  - to sustainable development

Fig. 3.20. The Innovation System to Support Structural Deficits and Changes of Paradigms ©IFF/IPA [3.33]

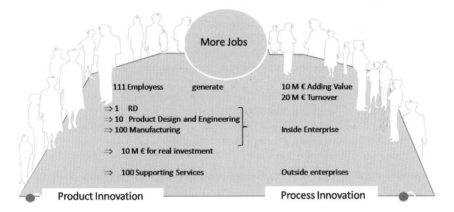

Fig. 3.21. Model Calculation for the Investment in RTD ©IFF/IPA [3.34]

Enterprises in the medium technologies spend 7 to 12% of their adding-value for RTD and the usual investment in workplaces is about 100.000€ for a human workplace. Based on this average data, a model calculation may be made, to find out the investment in RTD and workplaces which is necessary for the investment for generating Adding-Value in the manufacturing industries. The following example shows the impact of RTD on more jobs (see figure 3.21).

Investment in one employee for RTD creates usually ten jobs for design and engineering, 100 jobs in manufacturing and 100 jobs outside of enterprises in the public and private service.

Companies have to invest 10 M€ in factory equipment for these 100 jobs in manufacturing. They need a market volume, which is usually the double of the adding-value to cover the costs of material input and expenses for energy, external consultancy and taxes – in this case 20 M€ turnover.

Taking into account the rate of unemployment in the European countries, it becomes evident which great efforts have to be made for RTD and how strong the impact on employment and jobs can be. It is evident that research programmes for manufacturing must be based on the requirements of industries to reach the objectives of high-adding-value and sustainability.

References

[3.1]   Yasunaga, Y.: Eco-innovation & Future Actions with OECD. In: Japan Ministry of Economy Trade and Industry OECD workshop on Sustainable Manufacturing. Production and Competitiveness / Copenhagen (2007)

[3.2]   Rui, Z.: Changes in China's Five-Year Plans' Economic Focus (2005)

[3.3]   US department of Commerce: Manufacturing in America. A Comprehensive Strategy to Address the Challenges to U.S. Manufacturers (2004)

[3.4]   Working Document for the ManuFuture 2003 conference (2003)

[3.5]   ManuFuture Conference Milan (IT) (2003)

[3.6]   ManuFuture Conference Enchede (NL) (2004)

[3.7]   ManuFuture Conference Derby (UK) (2005)

[3.8]   ManuFuture Conference Tampere (Fin) (2006)

[3.9]   ManuFuture Conference Porto (PT) (2007)

[3.10] Jovane, F.: Fostering the K-based Industrial revolution. In: The European ManuFuture Initiative. 39th CIRP International Seminar on Manufacturing System, Ljubljana (June 7 -9, 2006)

[3.11] Report of the High level Group: ManuFuture Vision for 2020 – Assuring the Future of Manufacturing in Europe (November 2004)

[3.12] Report of the High level Group: ManuFuture Strategic Research agenda. Assuring the Future of Manufacturing in Europe (September 2006)

[3.13] EuroShoe Consortium: The Market for Customized Footwear in Europe Market Demand and Consumers' Preferences EuroShoe Project Report (March 2002)

[3.14] Westkämper, E.: Private Archive

[3.15] Malerba, F., Vonortas, N.S.: R%D Networks in the European ICT Industries. In: T2S Annual Meetings, Atlanta, Georgia (September 28, 2006)

[3.16] Specific Project no. 033416 LEADERSHIP - Leading European RTD Sustained High Value Innovative Production for Manufuture - Specific Support Action - Priority 3 – NMP

[3.17] Commission Staff Working Document - Accompanying the Green Paper 'The European Research Area: New Perspectives' (2007)

[3.18] European Commission - Green Paper 'The European Research Area: New Perspectives'. COM, p. 161 (2007)

[3.19] Final Results of the Public Consultation on the Green Paper "The European Research Area: New Perspectives" (April 2008)

[3.20] http://www.consilium.europa.eu/cms3_fo/ showPage.asp?i=1479&lang=DE&mode=g

[3.21] OECD Science, Technology and Industry Outlook 2006. Oecd (2006)

[3.22] Eurostat: Science, technology and innovation in Europe (2006)
[3.23] European Innovation Scoreboard 2006 Comparative Analysis Of Innovation Perform-
         ance (2006)
[3.24] Eurostat Yearbook (2004)
[3.25] European Union: The Treaty of Amsterdam – Amending the treaty of European Union
         (October 2, 1997),
         http://www.eurotreaties.com/amsterdamtreaty.pdf
[3.26] http://www.Cordis.eu
[3.27] Italian Chairmanship 07-08 – Eureka Strategy. Supporting European, RTD&I-based,
         competitive and sustainable industrial response to the global challenge. Working
         document (June 2007)
[3.28] http://www.eureka.be
[3.29] Borg, L.: The EUREKA Initiative – IGLO, Brussels (May 6, 2008)
[3.30] Mihelic, A.: EUROSTARS, New era for EUREKA (2007)
[3.31] Slovenian Chairmanship 07-08 - EUREKA: A Proactive Response to Global Chal-
         lenges in Industrial Research Ljubljana Strategy Paper 2008–2013 (April 2008)
[3.32]–[3.34]   Westkämper, E.: Private Archive

4

Research, Technology and Development for Manufacturing

The ManuFuture Support Group
E. Bessey, R. Bueno, C. Decubber, E. Chlebus, D. Goericke, R. Groothedde,
Ch. Hanisch, F. Jovane, J. Mendonca, A. Paci, E. Westkämper, and D. Williams

In this chapter manufacturing strategies are analysed. Following the European Way to Competitive Sustainable Manufacturing (CSM), they are analysed in terms of visions, concepts and actions to reach long-term, as well as medium term, goals and targets. The goals are referring to products and services, new business models, lean efficient enterprise processes, new ways of working. Innovating manufacturing engineering, from adaptive to reconfigurable manufacturing, to knowledge-based factories as products, to new Taylorism, to networking in manufacturing is analysed as well as digital manufacturing engineering.

The challenge of advanced Industrial Engineering, emerging manufacturing technologies and technologies beyond limits are also covered. The enabler role of manufacturing industries is underlined. This ManuFuture road is a contribution to support industrial strategic planning and work programmes for public trans-sectorial collaborative research. The visions, goals and targets follow the requirements for Competitive Sustainable Manufacturing (CSM).

4.1 The European Way to Competitive Sustainable Manufacturing (CSM)

Manufacturing – understood as a complex socio-technical system – is fighting for competitiveness and sustainability. Research can contribute massively to the ambitious ManuFuture goals by innovating in technology and management in all fields of operations. Innovative methodologies and technologies will change the traditional structure and influence enterprise strategies, product development and processes.

Manufacturing Strategies

Manufacturing Strategies are visions, concepts and actions to reach long-term goals and targets.

Manufacturing Strategies include partial strategies of products, markets and sales, networking of manufacturing and processes. All together they are the core of enterprise strategic development.

The ManuFuture Road is a contribution to support industrial strategic planning and work programmes for public trans-sectorial collaborative research. The visions, goals and targets follow the requirements for Competitive Sustainable Manufacturing (CSM).

Researchers and industries have to develop strategies for manufacturing taking into account the economic potential (lead markets) and the investment in product design and manufacturing systems.

It is necessary to join partial manufacturing strategies with the enterprise strategies and see investment in RTD as a strategic part of the Competitive Sustainable Manufacturing.

The reference model of manufacturing units in enterprises illustrates the functional elements of Adding-Value (figure 4.1). Direct and indirect processes are linked in the material and information flow. People work with machines, ICT systems and manually. The structure is highly dynamic and has to be adapted under the influence of external factors. RTD actions influence the system.

Figure 4.1 shows the functionalities of manufacturing. The upper area is characterised by business processes for management, authorities, engineering and planning. The lower level processes are the technical and logistic processes and chains of operations in the material flow. They are supported by peripheral functions.

The main question for RTD is which actions should be taken to reach the ambitious goals. Research for manufacturing must be oriented towards application and should have the potential for transformation to industrial practice. There are not only some specific technologies of highest importance but manifold opportunities. In summary, all actions to take should be contributions on the way of CSD manufacturing in Europe: The European Production System (figure 4.2).

Production systems represent the recognised and well-accepted frame for the management of manufacturing. They comprise the set of models, methodologies, technologies and tools for efficient organisation, engineering, planning, production and operations of products and processes.

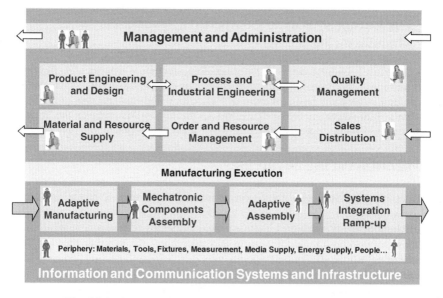

Fig. 4.1. Reference Model of Manufacturing Units ©IFF/IPA [4.1]

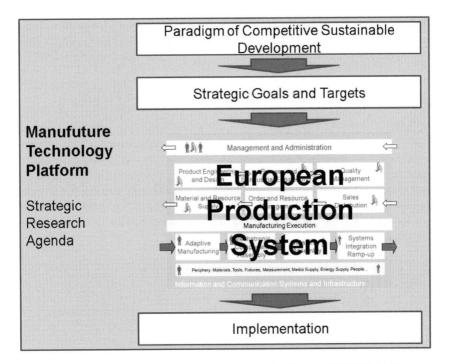

Fig. 4.2. The Way towards a European Production System ©IFF/IPA [4.2]

An accepted and available collection of about 80 methods is used to systemise the management of manufacturing processes. They include specific aspects like quality management, management of orders in logistic chains, planning of processes, resources or optimisation of time and cost. Many of them are implemented by using the modern Information and Communication Technologies (ICT) and tools. The basic foundations of these methodologies are partly old, as time or cost management or the optimisation of workplaces by ergonomic basics. Systematisation by use of basic methods brings manufacturing enterprises advantages in cost and value per capital, estimated at about 30%.

Behind these traditional paradigms, rationalisation, analysis of human work, division of labour, equal and incentive payment can be mentioned. They form the methodologies for lean management. These methodologies have to be oriented towards development and implementation of the so-called European Production System which is characterised by the culture of work, responsibility for sustainability (resources, management) and economic efficiency. Solutions are mainly influenced by permanent adaptation to changing conditions, activation of the potentials of global communication, global standards in logistics and technologies as well as global competition.

Therefore, the set of methodologies has to be extended and advanced towards the leadership of European Manufacturing in efficiency and human work. This can only be realised by interdisciplinary research and development of basics for a holistic view and the knowledge of a community. Interdisciplinarity is an essential aspect, by synchronising and harmonising economy, ecology, technology, communication, social areas and cultural aspects of work of millions of people. The European Production System is intended to become the standard of production around the world.

4.2 From Vision to Action: ManuFuture Action Fields

The stakeholders of the ManuFuture strategic role defined five pillars for CSD. Taking into account the variety of information and communication technologies in relation to industrial sectors and knowledge areas, the field of Information and Communication Technologies (ICT) received an enabling position in the strategic development. Figure 4.3 shows the main fields of actions towards the CSD oriented European production system for the sectors of capital intensive goods and consumer goods. Innovations in all of these action fields are contributions to European leadership in manufacturing.

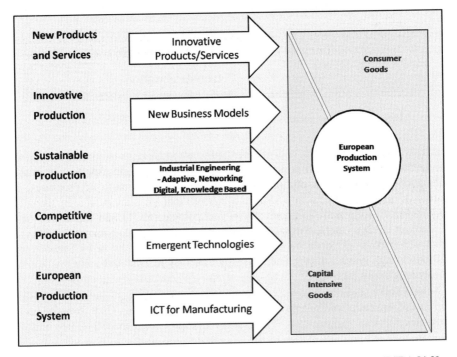

Fig. 4.3. Action Fields for Boosting Manufacturing in Leading Positions ©IFF/IPA [4.3]

The diffusion of ICT has reached nearly all workplaces in Europe. It becomes an enabler technology for manufacturing. But ICT technologies influence even the private way of life. Basic innovations and basic research for ICT will find its way into the area of manufacturing the industrial market and are even drivers of innovation in manufacturing.

The objectives follow the generic model of ManuFuture and objectives of High-Adding-Value and Sustainable Manufacturing. Strategic goals, targets and main visions elaborated by the ManuFuture actors follow in the next chapters. They can give a vision, which way manufacturing development should follow.

4.3 Strategic Goals and Objectives

The strategic goals, targets and visions of the action fields – presented in this chapter – have been developed by the ManuFuture platform.

The participants of the ManuFuture Conference 2004 in Derby, UK, discussed them and agreed in all points. Strategic goals and objectives follow the ManuFuture vision towards 2020. They may change when new aspects and influences will occur. Considering the life time of technical products and factories, industries have now to take the potential of RTD into account when defining their strategy.

Additionally the time between formulating RTD objectives and start of research in public funding systems is even an influencing factor on the road to completion and sustainability of the manufacturing industries. The following sections describe the strategic goals and targets for the European way to CSD in the main fields of actions and the key research areas.

4.4 Innovative Products and Services

Strategic Goal No. 1

European industries achieve business leadership through continuous innovation of new products. The concept of a product encompasses components, consumer goods and capital goods – and extends it to the provision of performance optimisation of entire production facilities. High-Adding-Value and superior quality result from the exploitation of world-leading developments of European RTD's, by enabling technologies such as innovative materials, nanotechnologies, ICT and mechatronics. The focus increasingly shifts from physical delivery of product to sophisticated provision of product-based functions and services.

Product innovation takes place in a complex network of actors, involving customers and their requirements, product engineering, the enterprise structures and physical manufacturing facilities. They are assembled together in order to realise the product/service concept and to deliver it to the customer (figure 4.4).

Visions for product innovations are driven by the idea of implementing technical intelligence in products and their integration in a global management system for service and life cycle optimisation.

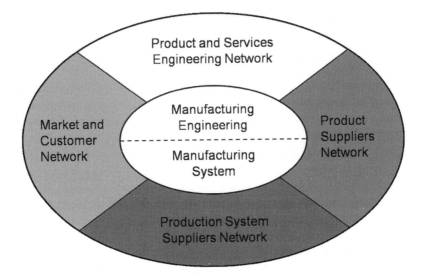

Fig. 4.4. Networking for the Development of New Added-Value Products (©IFF/IPA) [4.4]

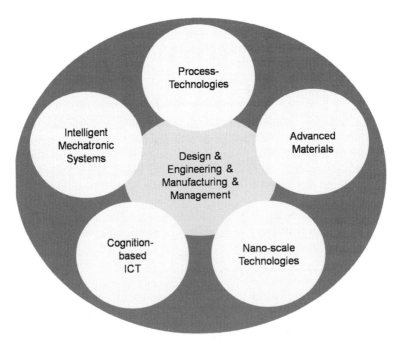

Fig. 4.5. Drivers for Intelligent Technical Products (©IFF/IPA) [4.5]

4.4.1 Intelligent Products

Intelligent products are characterised by the implementation of sensor-actor systems which make internal automation possible. Monitoring and supervising the influencing environment, learning functions, stabilising processes in critical situations, compensation of deviations and control with cognition-based on physical, chemical laws and phenomena are elements of future intelligent products. They allow the optimisation of usage.

Innovation drivers of such future technical products are shown in figure 4.5. The knowledge of processes which is the necessary knowledge base for control and optimisation of technical solutions follows usually cost and time intensive technical experiments.

There is a large field for basic and application research, to find out the basics of processes. If it is possible to understand such basics by scientific research and to develop process models, time and costs of experiments in early phases of the product design can be reduced.

Embedding electronics and sensors into the technical devices and components (intelligent mechatronics), higher performance and capable machines can be realised. Even advanced materials (e.g. adaptive or functional materials) and cognition systems or nano scale measurement in a product's technical system offer high potential for product innovation and new generations in all industrial sectors.

Design is the process which gives a physical shape to products that meet people's needs and desires. Apart from aesthetics, design can significantly contribute to utility value, and is often a decisive factor when choosing between different options. European industrial design contributes by adding value to products through functionality and aesthetics. Although design is already a strong point of many European products, the EU needs to leverage the strengths of its talent pool to a greater effect, since other countries are not behind the EU Member countries in technological innovation. Product design is the activity concerning the products system: the integrated body of products, services and communication strategies that either an actor or networks of actors (companies, institutions, non-profit organisations, etc.) conceive and develop as well as to obtain a set of specific strategic results.

This is based on requirements such as:

- response to life cycle processes and contextual conditions,
- compliance with competitiveness and sustainability goals, while pursuing the added-value approach; and will require:
  - the acquisition of enabling technologies covering architecture and components, structural and functional materials, and processing,
  - the increased incorporation of new technologies as they emerge from research,
- developments in enabling technologies, such as holistic user-centred design, innovative materials (smart materials, intelligent and adaptive structures), nanotechnologies, ICT and mechatronics give almost limitless possibilities to develop new products and rapid manufacturing or add functionality to existing product concepts. European industry must have immediate and future access to these technologies and to the tools for incorporating them into products design.

The market increasingly demands customised products, yet available with short delivery times. Fast decision making or rapid management need major attention to decrease the time-to-market and delivery times. Furthermore rapid manufacturing technologies will provide new means for manufacturing customised products. It is essential that European companies are able to understand and satisfy the needs of customers, regardless of their geographical location. The business focus will increasingly shift from only designing and selling physical products, to selling a system of products and services (described as 'product/services' or 'extended products') that are jointly capable of fulfilling specific users demands. This concept is equally valid for the products and machines used for manufacturing.

4.4.2 Services in the Life Cycle

Through life cycle orientation and the provision of product/services, European companies will gain:

- more opportunities for innovation and market development,
- more and longer term customer relationships,
- better feedback from consumers.

Product/services will offer greater satisfaction of customers' needs, reduce total life cycle costs and avoid problems associated with the conventional purchase, using maintenance and an eventual replacement of goods.

Product/services also have the potential to improve sustainability, compared with the recent buy-use-dispose products.

4.5 New Business Models

Strategic Goal No. 2

The continuing pressure of globalisation and changes in the structure of industries requires European manufacturing businesses to transform. Businesses must rapidly form networks of complementary capabilities to respond to market opportunities. An enlarged Europe enhances the opportunities for businesses to remain competitive. New generations of manufacturing enterprises must be created. A new, more networked and entrepreneurial, industrial mindset must underpin innovation and transformational changes in European manufacturing. The identification, promotion and application of new business models, methods and information tools will enable the growth of new businesses and will allow existing industries to sustain global competitiveness from a base within Europe. Enhanced capabilities that bridge entrepreneurship and technology management must develop new industries which use emerging science to meet emerging market demands.

As stated by Michael Porter "The business model is at the core of the competitive response of the firm to the market. A business model outlines how a company generates revenues with reference to the structure of its value chain and its interaction with the industry value system" [4.6].

Strategic business models are often framed in response to particular competitive circumstances, such as the length of a product's life cycle, or the length of the sector life cycle and its maturity. However, the underlying drivers are the same for all businesses. They are seeking to maximise their added value; profit from a differentiated capability; devolve risk; minimise exposure, headcount and capital expenditure; and thus optimise value for shareholders and other key stakeholders.

Visible changes in business models over recent years have included:

- a transition from products to services,
- the reduction of vertical integration in large businesses,
- a diffusion of intellectual property across companies and national borders,
- an increase in the importance of networks of small businesses, working in open collaborations to form a value system.

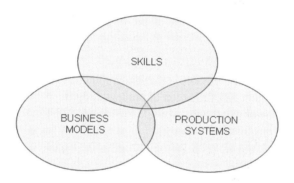

Fig. 4.6. Business Models for Future Manufacturing [4.7] © Loughborough University

The environmental perspective will modify business models by emphasising the whole product life cycle. Some national traditions will be challenged by business models arising solely from an Anglo-Saxon economic perspective. There are unique opportunities to build new brands and product-led businesses based on other European cultures, ethical traditions and aesthetics. Figure 4.6 shows that the enterprise acts to integrate three key elements: the production system which makes the physical product; the business model that matches the product with the market and determines how the enterprise generates revenue; and the skills and capabilities within the business, necessary for success.

With the continuing globalisation of manufacturing and the political integration of an enlarged EU, there is a need for the European manufacturing industry to enhance its business models, to:

- identify and exploit new opportunities for maximising value in the product life cycle,

- embrace global change in industry and business structures and their value systems,
- establish partnerships for economically sustainable manufacturing in an enlarged Europe that works with the rest of the world,
- understand how to realise and supply manufacturing and other technology/knowledge-based services,
- embrace innovation and entrepreneurship as the route towards a successful and secure business growth,
- recognise the product opportunities that emerge from markets and new global sciences, offering the growth of manufacturing businesses.

Lean, efficient enterprises

Businesses will develop efficient processes to create, manage and control the entire production chain and life cycle of products. RTD will address the development of intelligent controls, expert systems, improved and supply chain management; while the exploitation of emerging technologies will radically reduce the cost and time of designing, manufacturing, delivering, and supporting products.

RTD goals are:

- Development of design-for-manufacturing tools and techniques to facilitate the introduction of innovative concepts and new devices, thus shortening the distance between research lab and commercial fabrication;
- development of new basic models of processes and simulation techniques, integrating cognition, learning and validation of product designs;
- improvement of work conditions towards virtual and organised methodologies by using a codified information exchange;
- efficient specialisation of work procedures and specifications based on scientific results; optimised transfer and codification of SMEs internal know-how.

New ways of working

Rapid formations of open networks in both traditional and emerging sectors will bring significant increases in capability, profitability and productivity for all European businesses. Creation of environmentally benign product-based service companies will create a net increase in employment.

With better understanding of the innovation and technology management processes, continuous innovation and entrepreneurship will become embedded in the enterprise culture.

The transformation from traditional to knowledge-based operations could, for example, give small machine tool makers the capability to supply turnkey production services on customer sites. Furthermore, whole new manufacturing sectors will arise, based on new science or emerging markets.

Key research and intervention areas

To promote transformational changes in European manufacturing industries and the firms within, the education of industries to increase capabilities and human factors, the identification and promotion of necessary changes and researching and applying tools and techniques that enable changes, are necessary.

Co-operative research will lead to a better understanding and communication of evolving networked business models. It will determine how business strategy choices lead to enterprises and value network designs.

The production of pre-competitive demonstrators will facilitate the collaboration in networks to develop continuous learning and change of enterprise designs. Building secure and successful businesses in an open, highly networked and interdependent economy will require definitions of the most appropriate mechanisms for technology and product development, and for market and investment risk analysis.

Methodologies, measures and/or information tools must be developed to determine the fitness-for-purpose of business models, with respect to the requirements and life cycles of often complex product/service systems. An essential ability will be to exploit the whole life cycle of products, while minimising the environmental impact of their delivery.

Necessary operational improvements for the rapid implementation of collaborations in volatile networks include the management of IPR, decision processes, productivity, risk management, information & process integration and value retention across business interfaces.

Product and process technologies must be designed to allow the embedding of IP, thus safeguarding the interests of the originators. Equally important will be to understand the contractual and technical requirements for selling productions and other knowledge-based services. Establishing, how businesses in the enlarged Europe can work together, to deliver mutual benefit, entails awareness of the regional balance of finance, design and manufacturing.

The goal will be to increase business and economic growth, efficiency and local added-value, while retaining the manufacturing advantage factor of an enlarged Europe. Methods to identify new business opportunities from the early stage of science are required. Determining the requirements for improvements to the innovation process and foster entrepreneurship in science-based industries, likely to produce significant manufacturing value-adding, is also important.

Needed key skills will be found in networked lean new products and process introductions, project and supply chain management, and procurement. Vocational education and training systems will be required to equip people for working in networks.

Those capable of managing and financing innovation under high uncertainty and as a partnership activity will be highly valued. At the same time, instilling the need for ethical practices will be essential to uphold Europe's societal standards.

4.6 Innovating Manufacturing Engineering

A successful knowledge-based European manufacturing industry will be adaptive, networked and operating with digital tools and systems in a knowledge environment. The main tasks of Industrial Engineering will be the implementation of new technologies and innovation into operation and keeping enterprises on highest level of performance. Continuous adaptation in all levels of manufacturing is the key for completion and sustainability.

Strategic Goal No. 3

Manufacturing engineering is a strategic methodology, used to develop enabling technologies for planning, design, optimisation, adaptation, reconfiguration and recycling. Manufacturing engineering takes a holistic approach that includes the engineering of the enterprise structure, the development of the organisation, the design and process engineering and the tools and systems for high engineering efficiency. The agility of manufacturing engineering enhances the strategic development of enterprises. Manufacturing engineering is hence the "glue" for constructing factories, networks, all technical elements, equipment and IT systems for manufacturing, including services. Traditional factories have seen dramatic improvements in efficiency and changes in working methods brought about by the introduction of automation and control systems based on digital technology. With most influencing factors in a constant, and even turbulent state of flux, the next step is to progress towards what can be described as the 'virtual factory' of the future. This will require a European platform for digital manufacturing engineering, having the capability to create, use and maintain a dynamic system of networks, in which all actors contribute and add value, without the constraints of physical co-location or rigid partnerships.

4.6.1 Adaptive Manufacturing

Adaptive manufacturing focuses on agility and anticipation to permit flexible, small-scale or even single-batch-oriented production, through the integration of affordable intelligent technologies and process controls, for achieving optimal efficiency in which the human experience can be properly formalised and integrated.

Adaptive manufacturing responds automatically to changes into the operating environment. It integrates innovative processes as well as the capability to continuously change the structure of the socio-technical system. It integrates knowledge into the technical systems to control the processes at high-performance levels through intelligent combinations (such as intelligent mechatronics), and handles the transfer of manufacturing know-how into totally new manufacturing-related methods and systems. It embraces manufacturing systems and equipment, incorporating automation and robotics, cognitive information processing, signal processing and production control by high-speed information and communication systems.

Strategic Target No. 1

Development, supply and management of adaptive, digital and knowledge-based factories are major strategic priorities for European manufacturing enterprises in all sectors. Global leadership in the delivery of factories including all the elements – buildings, machines, systems, tools, training and other services for optimally efficient production – is an imperative objective.

'Manufacturing/enterprise engineering' simultaneously addresses all inter-related aspects of product life cycles, from design to disposal/recycling. Engineers who design products, processes and enterprises work with knowledge-based IT tools, operating in networks with standardised platforms.

Adding value by leveraging traditional European strengths, e.g. in engineering and design, while enhancing functionality by the incorporation of new technologies, differentiate products from those of the competition.

Industrial enterprises must re-examine their organisational structures and basic activities to accommodate the changes foreseen in manufacturing processes in a fully functional new industrial engineering environment.

For European industries playing a leading role at a global level, based on the RTD-to-market value chain, the factory itself has to be approached as a new and complex type of product with a long-life cycle, but able to adapt continuously to the needs and requirements of markets and economic efficiency: Operating every day on the best economic point.

New complex products and processes, together with production systems, will be developed through efficient reuse of technical/scientific, business, and process knowledge to make accurate decisions.

Shared knowledge of material and process properties and their interactions will support optimised process designs and a total understanding of complex transformations and interactions at micro and macro levels.

Creating new markets/industries, or gaining share in existing markets, will come through radical science-based product innovations (comparable in impact to e.g. walkman, cell phone, jet engine, semiconductor, laser, etc.).

Enterprises will seamlessly interconnect between their internal functions, and with external partners and stakeholders. This will permit operations of open, complex, distributed engineering and supply chains and extended enterprises, which function as integrated entities. Networking will improve the efficiency of information exchanges in terms of inventory levels and production/delivery schedules. The research focus will lie upon:

- adaptive networks, systems, machines, intelligent controls & functional elements and partial autonomy;
- interoperability and ongoing standardisation efforts that benefit from Europe's 'open' approach;
- integrated and product life cycle-oriented ICT infrastructures, providing the means for secure and reliable communication, in terms of both space and time; methods

and tools for knowledge management and collaborative environments related to either products, processes and markets.

Reconfigurable manufacturing

Rapid and adaptive design, production and delivery of highly customised goods will establish closer coordination between demand and supply sides. Continuous change will demand improvement in modelling and simulation of new complex phenomena (complexity, uncertainty management, multi-domain support). Enterprises will create networks of virtual factories, achieving reduced time-to-market, reduced order quantities, mass and extreme customisation, just-in-time production and reduced need to transport components and products. New paradigms such as ambient intelligence will facilitate the integration and adaptation of people and manufacturing devices. Research must determine optimal enterprise configurations and management of production and networks to improve the flexibility of all the manufacturing chains.

4.6.2   Knowledge-Based Factories as Products

Strategic Target No. 2

Factories can be regarded as socio-technical systems; they are capital intensive, complex and long-life products, operating through complex relationships between the material value chain and information chains, involving technical and human elements. In contrast with other complex products, factories have an overall system architecture enabling continuous adaptation to the needs of customised products, economic environment and objectives. Just as for other complex products, knowledge is the key to maximising the economic success and the dynamics of this socio-technical system.

Knowledge, which currently exists only implicitly within the skills of workers, technicians and engineers, is explicitly implemented in the systems of management, engineering and control of processes.

All processes inside and outside the factory are interlinked. The overall efficiency of the manufacturing network depends on the efficiency of each system element. European standards for knowledge-based manufacturing are able to compensate turbulent environmental influences by system based methodologies and intelligent technical solutions.

Knowledge-based manufacturing shows a deep understanding of the behaviour of machines, processes and systems. It will demand more research of simulation, meaning to integrate these inter-related aspects. Today, simulations are used for the engineering of logistics, machines and kinematics – and partly for processes. Future engineers will need multi-scale simulations, with high-performance computing and the ability to adapt to real or forecast system behaviour. New basic models of processes and simulation techniques must be developed, extended by automated planning and programming and possibly incorporating provision for cognition and learning features, as well as integrating diverse simulation aspects such as mechanics, control and process physics into unified models. Learning and reasoning will enable the system to cope with effects that exceed simulation capacities. Planning will make efficient use of simulation and process models. Knowledge-based manufacturing strives for a seamless integration of scientific, technical and organisational knowledge from all fields of production, such as process industries, advanced functional products, micro- and nano scale engineering, and intelligent mechatronics systems for high- performance design, engineering and production.

Learning is a central feature in knowledge-based manufacturing: learning by education, learning from experiments, learning from analysis of best practices, learning with scientific-based methodologies or learning with simulation machines (learning from the future). The main research objective is the development of concepts for factories, which are capable of adapting themselves continuously to the requirements and tasks of changing market requirements or changing product and production technologies. This continuous adaptation at all levels of the factory requires explicit and implicit knowledge and hierarchical system architectures, taking into account the complexity and synergetic work of networks.

The development of future knowledge-based factories requires research for adaptive structures and solutions which take into account the aspects of continuous changes by:

- management models and systems, following the objectives of self-organisation and self-optimisation,
- reconfigurable technical systems and integration of the processes into systems,

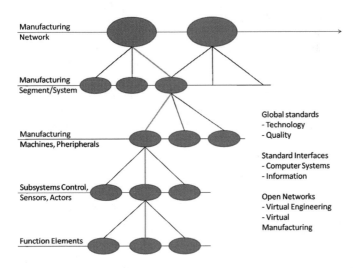

Fig. 4.7. Factories are Socio-technical Systems with Scales in the Hierarchy from Networks to Processes and Functional Elements (People and Machines). (©IFF/IPA) [4.8]

- technical intelligence with actuator and sensor integration by process control systems with process models and semi-autonomy and cooperation functionality elements, efficient in interacting in the human-machine interface,
- efficient networking in systems, based on standards and open system architectures.

The backbone of knowledge-based manufacturing is the information (digital) system. Therefore, it is important to develop a kind of a windows platform for manufacturing in networks which is scalable from networks to function elements (figure 4.7).

4.6.3 New Taylorism

Taylor defined the basic paradigm for manufacturing management more than 80 years ago. The Tayloristic organisation characterises the organisation model of nearly all manufacturing processes and systems and industrial manufacturers. They still use these methods to plan the operations in all areas of manufacturing.

Taylorism specifies the tasks of workers, based on elementary processes. The tasks are planned in detail by the usage of basic methodologies like MTM (Methods-Time-Measurement) or REFA (association for work design/work structure, industrial organisation and corporate development).

Process plans are based on highly detailed standard times. Global operating companies in the automotive and other sectors use this methodology to calculate, compare and to standardise processes world-wide.

The structural transformation from cost-driven to knowledge-based manufacturing of intelligent products and services will change the role of humans in all areas of industrial production. There is a trend to more indirect than direct work and there is a dominant trend to customers interests, which requires more flexibility and higher skills.

Strategic Target No. 3

Human involvement in manufacturing is essential for the effectiveness of manufacturing. People are not just the costly resource in our high wage area – they are the drivers of change and leading actors in manifold roles: workers, technicians, engineers, managers. Together they make the culture and climate and are the owners of the implicit know-ledge.

The motivation to maximise the potential of our "human capital" will be part of the strategy of competition and sustainability. For this target traditional paradigms have to be changed from old Taylorism to new and dynamic forms of knowledge-based manufacturing.

Additionally, people take care about optimisation and utilisation of production systems. Following Taylor's basics of dividing work and the new criteria of sustainability it is evident that wage systems and regulations of work have to be renewed. New

Taylor seeks for innovations in culture of management and work, motivation and engagement of workers with European social standards and methodologies for high dynamic and human oriented production.

This methodology is contradictory to the paradigm of a socio-technical system, characterised by knowledge-based manufacturing, manufacturing in networks or principles of self-organisation and self-optimisation. The concept of integration of knowledge into machines and systems is not compatible with detailed process planning.

As a consequence manufacturers need to adopt a new type of Taylorism, which takes into account the dynamic change and adaptation, the specific human skills and the requirements of cooperation in networks. This new European standard of manufacturing takes into account the social culture of Europe.

The factors accounting for the success of European manufacturing industries are mainly related to the great diversity and skills of personnel at all levels. Harnessing these abilities in the factories of the future will be vital to the economies of the Member States. It will be essential to adapt the structures as quickly as possible, aided by research into all aspects of human involvement in manufacturing.

Rapid evaluation of change under practical conditions, monitoring the success in meeting the demands of markets, and exchanging knowledge are the keys to growth and leadership.

RTD will be required to deliver:

- better understanding of new interdisciplinary fields,
- enhanced environmental impact assessment,
- interoperability and ongoing standardisation efforts that benefit Europe's 'open' approach,
- integrated and product life cycle-oriented ICT infrastructures that provide the means for a secure and reliable communication in terms of both space and time,
- methods and tools for IPR management and collaborative environments,
- empowerment of the workforce to operate in a rapidly evolving technical and organisational environment,
- formulating the European standard for work, maximising the potential contribution of all in factories.

4.6.4 Networking in Manufacturing

Networked and integrated manufacturing replaces the conventional linear sequencing of processes with complex manufacturing networks that often operate across multiple companies and countries. This mode of production makes it possible to insert processes and manufacturing systems into dynamic, Adding-Value co-operatives, and also to remove them when they are no longer needed.

At the lower levels of the factory structure, networked manufacturing requires the collaboration of system elements into the material and information chains. Networked manufacturing allows dynamic and Adding-Value co-operation in production focusing on synergy effects in the supply chains from customer to customer and networking in product/production engineering and supply of all resources needed for operations (figure 4.8). The efficiency of networking manufacturing depends on the skill of participants, the distances and culture of management.

4.6.5 Digital Manufacturing Engineering

Future manufacturing enterprises will leverage revolutionary technologies that radically change the way they design, engineer, manufacture, build and support products. The role of research will be to:

- integrate new technologies with currently applied standards and methodologies (non-disruptive approach),
- adapt new technologies according to users' needs, based on modelling at nano/micro/macro levels,
- develop engineering methodologies for the ubiquitous computer environment in product/process design, control and simulation.

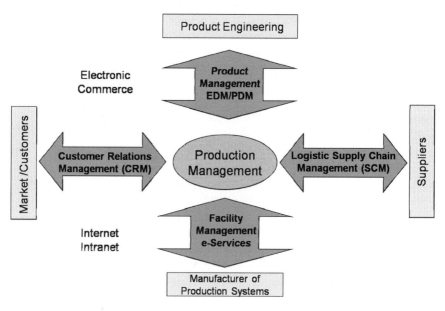

Fig. 4.8. Networking in Manufacturing ©IFF/IPA [4.9]

Key elements of this field of innovation are:

- holistic manufacturing systems technology,
- modular and configurable technologies,
- intelligent, flexible automation,
- real-time management systems,
- real-time digital factories,
- knowledge integration in control systems and embedded components,
- methodologies for management.

The trend will go towards the generation of robust design and planning systems, giving higher quality solutions in relation to the quantity of input information. Reaching this target demands:

- standardised models of product data and manufacturing resources,
- data management systems following the product and manufacturing life cycle,
- cognition-based tools and methodologies to minimise errors by dealing with uncertainties in automatically evaluating and elaborating solutions for complex systems.

In order to design and implement knowledge-based systems, great attention must be devoted to the development of 'self-learning' systems, able to use experience and histories of development processes in processing real-time data to extrapolate information or predict behaviour and to generate new knowledge by proposing several optional and solution alternatives. Knowledge-based systems will facilitate the rapid transfer of data across product-process domains and life cycle phases.

Digital manufacturing uses a wide range of planning tools, software and ICT to integrate new technologies into the design and operation of manufacturing processes and their corresponding production systems. Modelling and presentation tools allow the creation of a scalable virtual representation of an entire factory that includes all buildings, resources, machines, system equipments, labour staff and their skills. With the ability to simulate dynamic behaviour over the whole life cycle, planners and designers can gain dramatic time and cost savings in the construction of new facilities. By the same means, they are able to optimise reliability and minimise environmental impacts.

Strategic Target No. 4

Industrial Engineering (IE) is the key technology for pushing manufacturing towards competition and leadership. IE is the engine for knowledge-based manufacturing. IE has to accelerate the process from invention to application and industrialisation of manufacturing in emerging sectors. The new culture is knowledge-based, operating with the most advanced manufacturing tools of engineering, technologies, adopting business models and intensively integrating emerging technologies. Advanced Industrial Engineering opens the way to continuous optimisation of manufacturing. The factories themselves are regarded as complex, long-life products that operate with high-value technologies – continuously adapting to the customers' and market demands, and to the competitive technical and economic environment.

The concept of Industrial Engineering – the way that processes and production are organised in novel production patterns within factory units, able to respond flexibly to global demand – is the core of manufacturing development. It will be embedded in networks of product engineering, materials and component suppliers; in the network of manufacturing suppliers; and – in the future world of customised products – in the network of customers. This inevitably implies significant changes in manufacturing processes along the whole product value chain within the networked enterprise.

It will require both engineering competence and the tools to support engineers working in distributed and open networks. Here, their role will be to implement new technologies and to design manufacturing systems by employing an intimate blend of virtual and real world techniques.

The development of engineering infrastructures in open networks is thus the way to achieve leadership, high value addition and competitiveness. The better and faster the engineering processes, the greater the potential for success. As a consequence, there is a need for European industry to progress towards simultaneous development of product/services and processes within the enterprise/network organisations.

Process design will address processes throughout the whole product life cycle – design, production, distribution, use and maintenance, recycling – as well as the life cycles of individual processes, whose phases are design, implementation, use and maintenance, and reconfiguration. It will be based on such requirements as:

- interrelation with other product life cycle processes and the product itself,
- compliance with competitiveness and sustainability goals, while pursuing the added-value approach,
- an increasing incorporation of new technologies as they emerge from research.

Manufacturing Engineering has a holistic approach that includes the engineering of the enterprise structure, the development of the organisation, the design- and process-engineering and the tools and systems for a high engineering efficiency. At all manufacturing levels, the factory/manufacturing can be defined in its "current" and/or "future" states, under the so-called digital and respectively virtual representations. This relates to the employed digital methods and tools or simulation applications/systems, used to represent the static or the dynamic states.

Manufacturing systems engineering will address the development of machines and equipment, and the technical factory supply systems (energy, air, water, information) with integrated tools for design, analysis and simulation under real conditions.

The challenge of advanced Industrial Engineering

The research field of advanced Industrial Engineering arises from the need to enhance the Industrial Engineering concept with new models, methods and deployed tools for two main reasons. First, Industrial Engineering has to reflect the new way of operation, the so called "digital business". The "digital business" represents an enhanced mode of developing businesses, suitable for any industrial sector, which uses/employs the newest state-of-the-art information and communication technologies in any part/segment of the supply chain. The second reason resides in the requirement to approach (plan, realise and manage) the factory as a new kind of "transformable and adaptable" product.

Digital manufacturing brings dramatic cost and time savings in the implementation of new production facilities, through virtual representation of factories, buildings, resources, machines, systems, equipment, labour staff and their skills as well as permitting closer integration of product and process developments through modelling and simulation.

Leadership in this field will depend on the development of a European real-time platform resembling *Windows for Manufacturing* with well defined IT standards and the flexibility to allow sectors to apply their own specific solutions. Intense research on robotics and flexible manufacturing will be needed to deliver 'plug-and-produce' systems with integral automated services. This, in turn, will influence the way people work. A new methodology for the organisation of human labour, taking into account the European culture and work standards, will also be required. Digital manufacturing engineering as a key component of manufacturing engineering uses a wide range of engineering and planning tools, software applications, and information and communication technologies (ICT) for efficient and effective integration of new technologies into manufacturing processes.

The main area of research is the development of integrated tools for industrial engineering and adaptation of manufacturing taking into account the configurability of systems (figure 4.9). Digital manufacturing employs the: distributed data management, tools for process engineering, tools for presentation and graphic interfaces, participative, collaborative and networked engineering, interfaces to the reality. Digital manufacturing gives to the factory engineer the representation of the factory as it is today, that means the static image of the so-called digital factory/manufacturing.

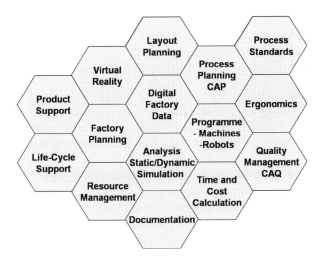

Fig. 4.9. Tasks and Areas of Digital Manufacturing Engineering and Components of the Digital Factory ©IFF/IPA [4.10]

Starting with the digital picture of the factory/manufacturing and by deploying the virtual manufacturing technologies consisting of simulation tools and specific applications/systems, the engineers deal with the factory and manufacturing processes in their dynamicity, by having the reflection of the "is" state in the future, the so-called virtual factory/manufacturing. At present, the developmental activities associated with the digital factory/manufacturing focuses on the planning of factories, production plants, new logistic systems, and of the manufacturing processes. Development and innovation of industrial products and processes is still experience-oriented.

Experiments and experience are the basics for reliability. In the knowledge-based industry, the 'costs of experience' – loss of productivity and time – can be reduced by modelling all the manufacturing processes in combination with (partial) automated planning and programming. Virtual Factories will integrate flexible supply chains for:

- engineering and designing products to match market needs,
- logistics, from customer orders to final delivery,
- consumable materials and waste treatment,
- factory machines,
- equipment and tools.

The constituent parts making up the virtual factory will be created from basic components, supporting transformability/changeability. Fast response at all scales, from individual processes to complete networks, will take place within a digital infrastructure, relying on a high level of know-ledge, and making extensive use of RTD for:

- the incorporation of functional and structural materials into machines, tools and other equipment,
- the implementation of ICT and cognition-based solutions for control and management of all processes,
- the application of micro and nanotechnologies,
- the enhancement of human/machine interfaces,
- the integration of methodologies from different disciplines for human work and management.

European manufacturing service industries have the potential to realise an open engineering platform, for which many different applications can be envisaged. The platform itself represents a future market, in addition to those for the products and factories it will produce.

The engineering platform can be modelled after product life cycle management systems. But there is a strong requirement to formulate the standards for the data management for all objects and elements of the factories. A specific objective is the real time management of any changes in the factories caused by wear, maintenance, set-up, end of life of machines and equipment. The digital engineering platform includes models of humans and workers (digital bodies).

A characteristic of next-generation manufacturing systems will be their evolvability. Here, the term is intended to indicate change that goes far beyond simple reconfigurability. As yet, this remains an unattained goal. Its realisation will depend on engineering, as the technology of the future, which can strengthen and speed the innovation process, support the progression from traditional to life cycle-oriented paradigms, and contribute to the science-based modelling of processes to realise science-based artefacts.

4.7 Emerging Manufacturing Technologies

Strategic Goal No. 4

RTD is opening new avenues to European leadership in manufacturing systems and technologies through the application of advanced solutions beyond the current state-of-the-art. European manufacturers set up world standards for factory equipment in all industrial sectors. Manufacturing technologies move consistently towards new levels of efficiency, and overcome existing technical limits by developing processes and intelligent machines. Technologies are able to handle and create more complex and unconventional materials, including those from biology. A holistic approach takes into consideration the converging nature of these sciences and exploits the potentials of process technologies, application of advanced materials, and implementation of intelligent mechatronics systems, cognition-based IT and downscaling of dimensions to the micro and nano scale.

Manufacturing is the integrator of basic technologies coming from natural sciences, material sciences and information sciences. Fundamental process knowledge is required for the realisation of new functionalities; in turn the application of the new functionalities in the manufacturing is required to make innovations happen. The objectives are the development of artefacts well beyond the current state-of-the art and the setting of new levels of performance standards and industry leadership.

The history of industrial technological developments can be traced in permanent innovations and solutions that have had advantages over former generations by being better, faster, cheaper, more convenient, and more flexible.

Even in conventional sectors like machining, the creativity of engineers has produced innovations, making it possible to step beyond the prevailing state of the art in processes such as:

- high-precision and ultra high-precision manufacturing,
- high-speed cutting, dry cutting,
- thixo-forming and casting,
- manufacturing of composite materials,
- rapid manufacturing,
- laser-assisted manufacturing processes.

Some of these innovations replaced traditional technologies; some introduced new product functionalities or reduced the consumption of energy and materials. Others brought new management methodologies, such as the 'lean manufacturing' concept originated in Japan.

Continuing to exploit emerging sciences and pursue specific objectives of RTD in emerging technologies will lead to:

- desirable new product characteristics,
- improved or new product functionalities of manufacturing technologies,

- intensive integration,
- new industries able to cope with global markets,
- radically enhanced efficiency in labour intensive manufacturing.

Many thousands public and private researchers in Europe – operating in the ManuFuture domain, within research institutes, universities and industries – can be engaged intensively in the continuous development of enabling technologies, blazing a new European trail towards implementation of the science market knowledge-production value chain.

They make up the most strategic asset of Europe: the knowledge base of an industry that can launch and sustain High-Adding-Value manufacturing engineering. Future advances will derive from fundamental knowledge of processes, and the successful incorporation of that knowledge into intelligent solutions. These will enable manufacturing systems to transcend current technical limitations in order to reach higher levels of performance and efficiency in terms of:

- new product properties (dimension, flexibility e.g. deformation, bending),
- production scalability/volumes,
- production speed, cost and quality,
- energy and materials consumption,
- operating precision (zero defect rates), and
- waste and pollution management.

The main objective is the acceleration of technological innovation in manufacturing by the development of high-end machines and systems. This includes the whole field of investment in assets for factories and the preparation for industrial manufacturing of products, based on new micro, nano and biotechnologies. Conventional manufacturing technologies have to be pushed to overcome today's technical limits in all respects.

4.7.1 Technologies Beyond Limits

Between state-of-the-art and physical limitations is a big field for innovations. In the past main criteria have been defined by cost orientation. Following CSD other criteria get higher priorities. Pursuing specific objectives, RTD can give rise to desirable new product characteristics, such as the reduction of weight – often leading to reduced energy consumption – through the use of new materials, new design and new bonding technologies. Many more product functionalities can be improved by manufacturing technologies.

Wear reduction, for instance, may be achieved through mechanical or tribological solutions. Other functions depend on optimisation of the reproducibility of production, e.g. for the manufacturing of high-precision photonic elements or in the development of fluid reactors, used in automation or medical/food production. Overcoming current manufacturing limitations by introducing new manufacturing sciences and technologies will help to drive innovation towards new industrial capabilities and the sustainable business of the future. To achieve these objectives, innovation processes centred on single competences will increasingly give way to a multidisciplinary convergence of: physics, mathematics, social sciences, biology, chemistry, medical sciences, etc.

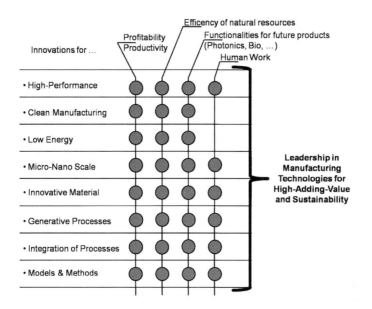

Fig. 4.10. Technologies beyond Borders –Implementation in Intelligent Manufacturing Systems and Artefacts (©IFF/IPA) [4.11]

In the medium term, added value is likely to come primarily from some of the most revolutionary technologies: machine intelligence (or artificial cognition), microelectronics, nanotechnology and biotechnology. However, the potential also exists to break new ground in the majority of traditional manufacturing technologies.

The contribution of extended technologies to environmental objectives is likely to be extremely high. A sustainable reduction of energy consumption by manufacturing processes and products over their whole life cycles would have a high impact on society. On the other hand, even modest advances will demand higher skill levels among workers and technicians in the factories.

An intensive engagement in emerging technologies through the cross sectional form of ManuFuture will be beneficial to the manufacturing industry as well as for more focused technology platforms. The ability to generate new knowledge and new capabilities increases with an early engagement in new technologies. On the other hand, potential and not foreseen "killer-applications" could possibly be identified in the manufacturing sector as follows (figure 4.10):

High performance: To overcome existing technical limits in manufacturing it is necessary to find the theoretical potential of all manufacturing processes and to activate it by the design of new machine generations and artefacts.

Micro scale technologies: Conventional technologies have their limits in μm dimensions. For high-performance micro manufacturing, technologies for machine internal measurements in nm dimensions are required. For this, the dimensions of sensors and actuators must be downscaled. Optical technologies for measurement have to be integrated in manufacturing systems. They will contribute to the industrial operation in the nano-scale dimension.

Clean technologies and reduction of environmental pollution address the existing process limitations. The objective is to fulfil all future regulations and reduce the contamination of air, water and humans by production technologies.

Generative technologies address the generation of usable forms, parts for rapid product development and manufacturing in series: shorter cycle times, repeatable quality and precision, lower costs. It even takes into account new solutions for the integration of sensors, actuators and functions in one step operations.

Energy and materials refer to the consumption of these resources by manufacturing processes. It is evident that both are increasing cost factors.

Methodologies for monitoring & management are part of holistic manufacturing systems. Operational losses are caused by the lack of appropriate methodologies for example in the field of quality management, inspection, diagnosis, learning from experiences, technology evaluation and many more.

Just taking into account the necessity of adding value by product- oriented services, new methodologies are required for efficient operations and management. One important field is the basic methodology of life cycle control of technical products.

4.7.2   Technical Intelligence for Efficient Processes and Usage of Resources

Scientific research offers possibilities to implement technical intelligence in manufacturing. Technical intelligence is realised by the integration of knowledge from research-based process models and practical experiences into products.

The vision of intelligent manufacturing is the following:

- research of processes defines the modes of operation using basic phenomena under the influence of dynamic parameter variation. It formulates basic process models for material transformation from raw material to parts and products, dynamic behaviour of equipment and tools and outside influencing parameters,
- multi-sensor systems supervise processes and statistic analyses of reproduced processes
- both are formulated in simulation systems and implemented in engineering tools as well as in the control units of processes.

Simulation needs scientific research in all technologies of manufacturing. This will contribute to the reduction of defects and the efficiency of technical systems. The basic concept is shown in figure 4.11.

Research is required in multi-sensor systems, modelling and simulation of processes and the integration in machines towards technical intelligence. Machines of future generations are able to reduce the consumption of resources and are a strong contribution for innovative manufacturing systems.

4.7.3   Flexible Automation

Manufacturing follows the development of flexible automation with a continuous innovation process. There is a high technical and economic potential in the future. The process of customisation in manufacturing requires higher flexibility of technical systems to change the operations and reduce the cost of retooling.

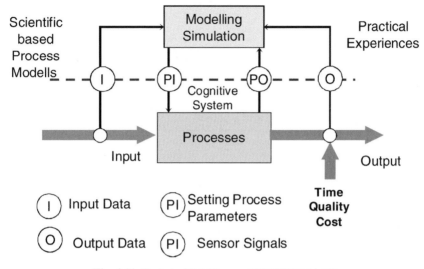

Fig. 4.11. Technical Intelligence (©IFF/IPA) [4.12]

Nearly all industrial sectors as well as emerging areas like the bio sectors need innovative solutions for the industrialisation of manufacturing. Core areas of this development are the manufacturers of machine elements, actuators and control systems.

Automation industries are a core of the future development and need research for the generation of new solutions for interfacing human-machine interactions and hybrid systems of the future.

Nearly all technical developments, management systems, administration, business and manufacturing processes need the support of ICT. There is nearly no working place in manufacturing without ICT support. Manufacturing is – outside of the private area – the leading market for ICT.

4.8 ICT for Manufacturing

Strategic Goal No. 5

Information and communication technologies (ICT) are driving the way to knowledge-based and intelligent manufacturing.

Manufacturing needs efficient tools for Management, Engineering and Operations with high integration in internal and external ICT architectures and applications. Open systems and embedding electronics enable the realisation of high integrated knowledge-based manufacturing. Standards, Modularisation and System Engineering will contribute to the changeability and transformability of manufacturing.

In the field of operations, adaptable manufacturing systems and robots support the targets of high efficiency and capability of processes. The enablers of high-performance are intelligent components with embedded electronics and sensors.

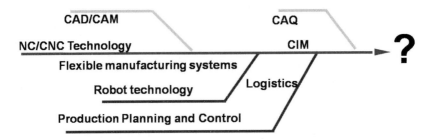

Fig. 4.12. CIM Evolution and Development ©IFF/IPA [4.13]

Figure 4.12 illustrates the former development of Computer Integrated Manufacturing (CIM) as a technical-oriented strategy of production in the last century. The CAD/CAM technologies, production planning and control and the process automation have been integrated into an overall architecture of computerised manufacturing. In the reality these strategies followed old and traditional organisational procedures. They were centralised and hierarchically configured. The central component was a common information and data base.

This system architecture can be characterised as technocratic because of its low flexibility for changing tasks and its total planning approach. But the main developments of current innovative ICT Systems follow the traditional CIM Vision. The vision of CIM in the last decades of the last century were driven by full integration of processes in open systems architectures (CIM-OSA).

Current manufacturing requires modular systems architectures, which are able to transform the principles of changeability and implement innovative systems for communication and visualisation. The high variety of processes and the need of efficient engineering and operation make it necessary to develop special applications and tools to support the processes and fasten the time of implementation.

4.8.1 Management Systems

Strategic Target No. 5

The development of European platforms for the processes of engineering, the management of orders and resources, dynamic administration, manufacturing execution, monitoring of processes is another strategic target. ICT systems for manufacturing have to be linked permanently to the reality and operate in real-time.

Simulation has to become a part of optimisation and adaptation. Simulation systems support "learning from the future" at all levels from networking down to the processes (multi- scale). The time scales cover ranges from μsec to long term.

The backbone of efficient ICT is represented by standards for data and information exchange. Europe needs a standardised ICT platform for manufacturing supporting the internal processes and the networking as base of future co-operative and knowledge-based European manufacturing.

Following these targets it seems to be possible to push European software enterprises the so-called ICT key players to worldwide leading positions. The common backbone of system architectures and a standardised platform of ICT inside and outside of factories give room for application specific tools and systems in the areas of product development, manufacturing engineering and management of resources.

Figure 4.13 illustrates the fields and hierarchies of manufacturing processes, structured according to the following four levels:

- management and administration,
- engineering of products and processes including the management of capacities and orders,
- manufacturing execution (cockpits),
- process control.

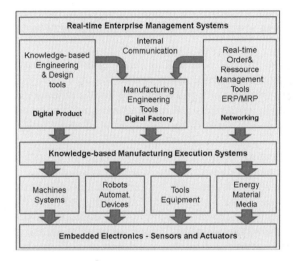

Fig. 4.13. Integration of Internal ICT for Manufacturing ©IFF/IPA [4.14]

The vision of long term development is the realisation of real time support with information and a distributed and open network including innovative technologies like wireless, ubiquitous computing, sensor integration, intelligent monitoring etc. The development of future ICT for manufacturing has to take care of the specific requirements of all these processes and specificities of companies. The engineering of software and the application into the existing infrastructure has to be made much more efficient and reliable. Software Engineering needs tools for application.

Companies' internal ICT infrastructures have to be separated from the external environment regarding the protection of knowledge. But on the other side it must be open for networking in the supply chains:

- customer – OEM – supplier over the life cycle of products,
- dislocated engineering and design in product development from idea to product,
- dislocated co-operation between manufactures of equipment and facility management over the life cycle of factories and their usage,
- management of products life cycle from birth to the end of life including maintenance, services, logistics & re-production and recycling.

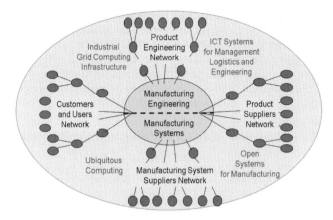

Fig. 4.14. ICT Infrastructure for Manufacturing ©IFF/IPA [4.15]

Figure 4.14 shows the vision of ICT as core element and infrastructure for European manufacturing. All involved actors can be linked by a high-performance ICT Network. It has the character of a grid computing system with standardised interfaces. In this network, software suppliers and manufacturing industries can operate on a high level of performance. The network has the potential of world leading in manufacturing support.

Ubiquitous computing, calm technologies, self-organising open networks, and location systems link mobile objects and management systems.

There is a need for basic research for European ICT standards. Data related to products and processes have a long-life and the change of basic data systems inside and outside of factories is extreme expensive. The definition of data systems has to take into account the requirements of lifelong documentation, the exchange between applications and the short life time and development of IT hardware and software. Fast innovations must be possible to adapt ICT permanently to changing requirements but the information basics are constant.

4.8.2   European ICT-Platform for Manufacturing (Grid Manufacturing)

Strategic Target No. 6

The development of a European platform for manufacturing is one of the most important topics for efficiency and competition of European manufacturing. It opens high synergies in the networking of industrial processes for co-operation and links the actors from customers to manufacturers and service companies. It is the platform for the exchange of information, fast transfer of knowledge from research to application and activation of losses in the usage of products.

Europe has the potential to realise a global network for manufacturing and can open a wide field of Adding-Value for the development of software applications and tools for manufacturing and management.

The ICT platform is even one of the enablers for activation of the potential towards the knowledge-based community. ICT requires RTD in the basics of communication technologies but also public investment in the infrastructure for industrial manufacturing.

Even the manifold applications offer an extreme spectrum of innovation by interdisciplinary RTD: manufacturing technology, management of manufacturing, economics, IT and social aspects.

4.9 Enabler Role of Manufacturing Industries

The sectors of the capital intensive industries play a key role in the implementation of CSD. Many sectors of the consumer industries and especially the manufacturers of mass production do not have the engineering competences for implementation. Especially the competences in industrial engineering are poor and insufficient.

The main role for innovating the production systems lies with the suppliers of methodological and technical solutions at the request of the consumer industries. Suppliers are: consulting, engineering, application-oriented research institutes, machine industries, software industries, training organisations etc. Here is the potential for development and implementation on a global level.

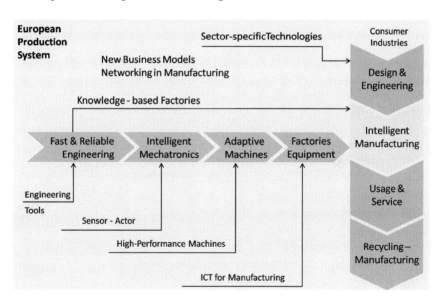

Fig. 4.15. Enabler Role of Manufacturing Industries ©IFF/IPA [4.16]

Figure 4.15 shows the way of innovations in the process chains over the life cycle. Driven by the basics and philosophy of a European production system (follows CSD strategic target and goals) and supported by innovative methodologies and technologies the process of application and diffusion can be accelerated for advantages of the whole area of manufacturing.

Strategic Target No. 7

Innovating production brings efficient and competitive manufacturing of all products: it functions through networks of OEMs with value-chain partners and suppliers of factory equipment/services.

The new culture is knowledge-based, operating with the most advanced manufacturing technologies, adopting new business models and intensively integrating emerging technologies.

Advanced engineering opens the way to new products. The factories themselves are regarded as complex, long-living products that operate with high-value technologies – continuously adapting to take account of customers' and market demands, and of the competitive technical and economic environment.

Innovations pushed by RTD in manufacturing are contributions to the development of the European economy. The strategic goals and objectives follow visions and activate the potential of competition and sustainability. Specific basic RTD areas like nano technologies, material technologies or ICT (cognitive) are extremely useful for the future but their implementation and transfer to Adding-Value has to be realised by engineers, who are able to implement these technologies in products, processes and management systems. Figure 4.16 shows the long-term road to implement basics into industrial solutions.

Following the strategic lines to Competitive Sustainable Manufacturing it seems to be possible to increase the leading position of European manufacturing in the world by implementing emergent technologies in Products and Processes. Engineering is one of the key technologies for fast application and transformation of Research into

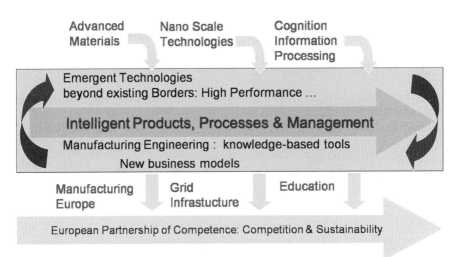

Fig. 4.16. On the Road to Competition and Sustainability ©IFF/IPA [4.17]

the practice. But even engagement of industrial enterprises and investors makes it happen that manufacturing can reach the goals and objectives. Governmental support is required from throughout Europe to push RTD towards Manufacturing, Development of the Infrastructure and Education of people. Actions on the way towards implementation are explained in the next chapter. Manufacturing follows the development of flexible automation with a continuous innovation process.

There is a high technical and economic potential in the future. The process of customisation in manufacturing requires higher flexibility of technical systems to change the operations and reduce the cost of retooling. Nearly all industrial sectors as well as emerging areas like the bio sectors need innovative solutions for the industrialisation of manufacturing.

Core areas of this development are the manufacturers of machine elements, actuators and control systems. Automation industries are a core of the future development and need research for the generation of new solutions for interfacing human-machine interactions and hybrid systems of the future.

Nearly all technical developments, management systems, administration, business and manufacturing processes need the support of ICT. There is nearly no working place in manufacturing without ICT support. Manufacturing is – outside of the private area – the leading market for ICT.

References

[4.1, 4.2]   Westkämper, E.: Private Archive
[4.3–4.5]   Westkämper, E.: Manufuture Strategic Research Agenda (2004)
[4.6]   Porter M.: The Competitive Advantage of Nations (1985)
[4.7]   Williams, D.: Archive
[4.8]   Westkämper, E., Zahn, E. (Hrsg.): Wandlungsfähige Produktionsunternehmen Das Stuttgarter Unternehmensmodell. Springer, Heidelberg (2008)
[4.9–4.17]   Westkämper, E.: Private Archive

5

The Proactive Initiative ManuFuture Roadmap

E. Westkämper

With contributions of the Leadership Roadmappers Group:
E. Carpanzano; C. Constantinescu, C. Decubber, P. Elsner, R. Groot-hedde,
B. Hellingrath, M. Höpf, H. Holezek, F. Jovane, H.-J. Korriath, S. Nollau,
A. Paci, H. Pflaum, D. Reh, S. Stender, W. Schäfer, C. Schaeffer,
D. Williams, M. Witthaupt and many others

A Roadmap for manufacturing research has been developed by the Leadership Consortium in 2006 and 2007 – it is based on the ManuFuture Strategic Research Agenda. The European manufacturing area has more than 23 industrial sectors. Several of them were involved in the process of the Leadership Consortium's roadmapping, supported by the EC-DG Nano- Materials and Processes (NMP). The result of this process was a set of more than 300 possible themes for research and a set of more than 100 proactive initiatives (see annex II). Priorities follow the general strategy of ManuFuture towards leadership in manufacturing through innovative solutions. Most of the proactive initiatives are of common interest to all manufacturing industries. Together they build the trans-sectorial roads for implementation of ManuFuture. They emphasise the global economy, in-cluding growth, jobs, environment and knowledge community.

5.1 Developing a Roadmap for ManuFuture

Investment in RTD for manufacturing is necessary to achieve the ambitious goals and targets of competition and sustainability in the wide range of industries in Europe. More than 85% of the proposed activities are of common interest to all industrial sectors. [5.1] The authors have summarised them into trans-sectorial Roadmaps, unified under a comprehensive approach representing the ManuFuture vision of European industrial transformation, the actions in these Roadmaps form the core of the ManuFuture work programme.

The roadmappers of Leadership found many specific actions in every field and discussed them with experts from industry. The actions have been synthesised into a readable Roadmap subsequently evaluated and classified in sectorial workshops. Each of these workshops started with the Vision and Strategic Research Agenda of ManuFuture. The results were discussed in the ManuFuture Conference 2006 in Tampere, Finland.

The priorities follow the main phases of technology life cycle, respectively innovation, development, diffusion and substitution. Experts from the manufacturing research community and industrial organisations analysed and formulated more than 300 possible RTD topics,'enabling technologies', and set up their priorities by taking

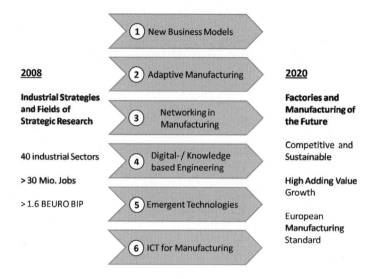

Fig. 5.1. Enabling Fields of Innovation – following CSD and Industrial priorities (© IFF/IPA) [5.2]

into account the technological and economic potential and the time scale for implementation. They all agree that the amount of financial resources spent for research by industries themselves and public funding in co-operative research projects and infrastructure has the most significant influence on the speed of development and the diffusion into practice. National, regional and sectorial priorities are now available and can be found on the national and regional platforms. There was a high correlation between different sectors which makes it possible to define a common trans-sectorial Roadmap for the 'European Road of Manufacturing' of the future. Inevitably, most of the proactive initiatives defined could be started immediately. However, given the current situation of many enterprises, dependencies and other strategic considerations it has been found necessary to order the actions in time.

5.1.1   Priorities for Industrial Implementation

Many enterprises are struggling to survive in the currently turbulent markets, some are leaders in markets and effectiveness. Others seek the future in services and emerging technologies.

The way to activate economic potential by research across the whole breadth of manufacturing is to take into account both sector-specific situations and trans-sectorial synergies. Thus industries can be pushed into leading positions and accelerate the diffusion process from basic research and innovation to application research. The Roadmap is driven by industrial/economic requirements and the need for the transformation of manufacturing towards CSD. It follows a strategy that:

- increases the efficiency of survival and transformation of enterprises to the requirements of customisation and sustainability,
- boosts the level of technologies of products and production towards global leadership,

- globalizes Europe as producer of factories and factory equipment (lead markets) with intelligent products and new business models,
- activates the potential of emerging technologies and develops solutions for emerging markets.

These industrial priorities take into account the problems of today's competition and the threats of migration. 60% of all companies do not have the resources for long-term strategic development. Their first interest is the short-term optimisation of their productivity and flexibility by implementation of lean management methodologies and an increased customer-orientation.

Many companies in the manufacturing area are leaders in the world market or have high market share. Their base is technical innovation and strong customer-orientation. They are known worldwide as high performing manufacturers and deliverers of high-quality technical products. Many are SMEs. They need research to maintain their position or to expand their global market share (see figure 5.2).

Many others have the competences to lead in technologies but fight against competitors from other economic areas. It is of utmost economic importance to support these competences and reduce the risks of emergent technologies by research.

The priorities developed take into account the risks of new technologies and the life cycle of innovation in emerging technologies. They are oriented towards the realistic market conditions in each sector.

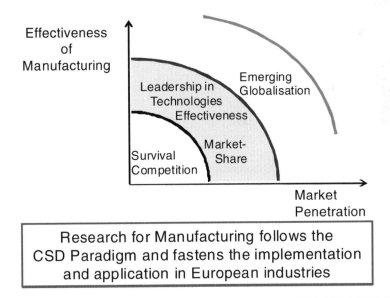

Fig. 5.2. The Road to Comprehensive Sustainable Development – CSD (© IFF/IPA) [5.3]

The implementation strategies envisaged are the following:

Competition and **Customisation (A)**, with high priority and immediate implementation, **Leadership (B)** in technologies and **Globalisation (C)**, aimed at high priority medium-term deployment and **Emerging (D)**, focusing on the long-term and creating

enabling technologies with the potential to accelerate innovation in emerging industrial sectors.

The economic pressures on manufacturers in Europe make it necessary to focus on
research as part of an overall enterprise strategy with the objectives of economic profitability in the global economy. Mass and series production in the low and medium
technology areas and for more conventional product technologies are now migrating
to regions with low costs. To push enterprises to higher profitability, it is necessary to
implement methodologies and management systems such as lean management, particularly taking into account the need to adapt these to different cultures or changing
production systems. Many SMEs do not have the potential to develop innovative production systems.

Europe has a long tradition of manufacturing technical products, based on inventions
and innovations: Manufacturing in Europe has a human orientation with a highly skilled
and qualified workforce, including scientifically-based social standards of work. The
"European Production System" has been challenged by the Far Eastern production system known as "Lean Manufacturing" or the "Toyota Production System" (TPS).

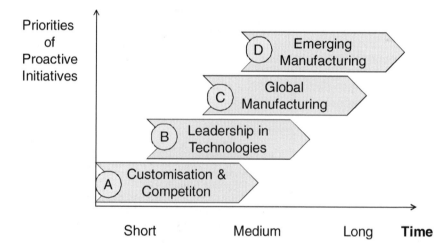

Fig. 5.3. Priorities Follow the Needs of the Current Situation and the Market Potential (©
IFF/IPA) [5.4]

Here the organisation is permanently fighting for efficiency concentrating on value
processes and logistics, simplification and standardisation, elimination of waste, loss
of time and quality and fast reaction to defects. Western automotive manufacturers
have implemented TPS, to improve efficiency and to remove disadvantages as a consequence of high costs of labour.

Production systems from the Far East are the default international standards for
manufacturing today. To overcome this standard with a European way of manufacturing, intensive and well focused research actions and developments of innovative
methods are necessary. With respect to time and priority, several research directions
are necessary (see figure 5.3):

Short-term orientation **and high priority** is the acceleration of the diffusion process for lean manufacturing knowledge into all sectors of manufacturing, through dynamic and customer-oriented strategies and innovative methodologies, thus supporting the manufacturing enterprises to survive or to stay competitive in the turbulent environment of global competition and high-wage economies.

Mid-term activities should focus in a number of complementary directions such as the definition of a European way of manufacturing, based on European cultural traditions, innovation and knowledge and oriented to sustainable manufacturing.

Implementation of the principles and new philosophy of 'New Taylorism' in all industrial sectors and integration of new industrial paradigms for manufacturing fitness, balancing reactivity and efficiency.

The **Long-term** orientation must be implementation of such a European production system as a standard and philosophy for knowledge-based manufacturing, adding global competitiveness and sustainability to the new industrial paradigm.

5.1.2 Economic Potential and Lead Markets

European manufacturers have disadvantages in labour costs and other influencing factors, such as working hours, against their competitors in developing regions. However, they have advantages in the skills and qualifications of their workforce as well as in the ready availability of the multiplicity of technologies needed for technical innovations. Therefore, it is necessary to follow roadmaps with an orientation to the creation of innovative products in all industrial areas and develop manufacturing technologies for innovative products.

The sectors of investment goods have the key role as enabler of this development. They produce products for the manufacturing of innovative products and make the highest contribution by allowing industrialised manufacturing of innovative products and in turn reduce costs by implementation of innovative manufacturing systems and technologies in all industrial areas.

They influence the performance, quality and time of processes, machines and factories. ManuFuture sees factories as complex systems and long-life products. It is evident that many innovations in manufacturing technologies, coming from more than 25 industrial sectors, define the efficiency of factories and influence the success of products worldwide.

The global market share of Europe in what is needed for manufacturing is around 25 %. This can be increased by one to two percent per year through the implementation of strategic proactive initiatives and fast diffusion to manufacturing workshops globally. Further, Europe has the potential to set industrial standards in a wide area and in the biggest markets. Lead market considerations and market potential have influenced the priorities of the Roadmaps.

5.2 ManuFuture Trans-sectorial Roadmaps

The ManuFuture work programme represents one of the main scientific outputs of the ManuFuture research agenda, aiming at identifying the relevant research topics corresponding to the ManuFuture Pillars and developing their trans-sectorial Roadmaps.

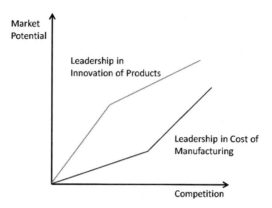

Fig. 5.4. Orientation to Fast Innovation (© IFF/IPA) [5.5]

Figure 5.5 presents the enabling technologies corresponding to the Manu-Future pillars, listed on the left side of the figure. The distribution and the employment of these technologies on a generic time scale and priorities are drafted on the X dimension of the picture.

Sector-specific actions with a focus on manufacturing have also been defined in co-operation with other European technology platforms. They are included in the SRA. The strategic fields of most of other platforms are concentrated on product technologies or emerging technologies. Therefore, this Roadmap is concentrated on enabling and enabler industries.

		Competition Customization ⟹	**Leadership** ⟹	**Globalisation** ⟹	**Emerging**
	New Business Models	Beyond Lean… Life Cycle Services Survival Strategies	European Production System Knowledge & Service	Real-Time Enterprises New Taylorism	Invest in R&D Enterpreneurship
adv. Industrial Engineering	**Adaptive Manufacturing**	Adaptive Automation Modular Products Configurable Systems	Adaptive Factories Real-Time Adaptation Adaptive Systems	Real-Time Factories Disruptive Factories	Knowledge-based Factories
	Networking in Manufacturing	Network Engineering Interoperable Networks Customisation	Manufacturing on Demand Networking Standards	Supply Chain Mgt.: - Real-Time - Global	Knowledge-based Order Management
	Digital Engineering	3D PLM and Tools Fast Engineering Digital Prototyping	Multi-Scale Simulation Digital Factory Material Engineering	Process standards Smart Factory Cognitive Simulation	Knowledge-based Engineering
	Emergent Technologies	Intelligent Products High Performance Energy Saving	Gen. Technologies Adaptive Materials Micro & Nanotechn.	Reliability Process Models and Simulation	In-Situ Process Control beyond Borders
	ICT for Manufacturing	Configuration Systems Embedded Systems	Multimodal Interfacing Software Engineering	Grid Manufacturing Ubiq. Computing	ICT Environment Manufacturing

Fig. 5.5. ManuFuture Enabling Technologies for the Next-generation Manufacturing and European Production Systems [5.6]

5.2.1 Proactive Initiatives for New Business Models

Current manufacturing systems, the typical instantiations of modern socio-technical systems called factories, have to solve highly complex tasks with increasing demands for adaptability, economic performance, maintainability, reliability, scalability and safety.

They operate in a turbulent environment. The 'next-generation European factory', when approached as a complex long-life product, has to be adapted continuously to the needs and requirements of the markets and economic efficiency. Furthermore, the factory will have to take into consideration more and more social responsibility and, in particular, environmental sustainability.

Based on these challenges, the need for the development and validation of new industrial models and strategies, known in short hand as "New Business Models" is more and more relevant for the purposes of the European industrial transformation.

A collection of enabling business and management models, methodologies, technologies and tools have been identified and then integrated , in order to support the implementation of the desired "Management of European Production Systems", into the following **ManuFuture Trans-sectorial Roadmap: New Business Models** (figure 5.6).

Graphically presented, the management system of European production represents the synergy of several major enabling business and management models, technologies and tools, structured in the following main clusters:

- European production systems solutions,
- European management systems,
- innovative management models,
- methodologies and tools and
- service and consumer-oriented business models.

The development of a European manufacturing solution and standard for a holistic production system is the objective of proactive initiatives to reach High-Adding-Value (HAV) and sustainability. These are issues of innovative research and technological development areas and are presented below from a technical point of view, the scientific objectives and potential results of each identified area or field are identified.

The Roadmap graphically presents several research directions on time dimension and priority scales.

Short term with high priority, innovative methodologies for supporting manufacturing enterprises to survive or to stay competitive in the turbulent environment of global and high wage economies, coming from two areas which are relevant: Strategies for transformation management, part of the cluster 'Management of European Factories' and survival production strategies, as a contribution to European manufacturing.

This is a dynamic socio-technical system which is operating in a turbulent environment, characterised by continuous changes at all levels, from networks of manufacturing systems to the factories, production systems, machines, components and technical processes. It is even necessary to develop and implement sets of methods for reduction of natural resources in the management system: Total energy and material management, management of hazardous goods and fluids.

Innovative, service and consumer-oriented enterprises

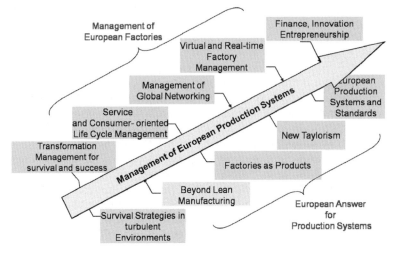

Fig. 5.6. ManuFuture Trans-sectorial Roadmap: New Business Models [5.7]

Medium term with high priority are solutions with the aim of enabling European factories to carry out service and consumer-oriented life cycle management and globally networked virtual and real-time factory management.

For European production system clusters, the research efforts have to be directed towards the investigation of beyond lean manufacturing, the employment of New Taylorism as a base of increasing the efficiency of people and manufacturing processes. Research efforts must be synchronised and harmonised to allow the implementation of factories as products, a new industrial paradigm and core business model. This includes the orientation towards sustainable manufacturing by concentration on economic, ecologic and social effectiveness of business. This offers new markets.

Long term, new business models, methodologies and tools have to be created to support the development of the concepts of and implementation of European production systems and standards. This will be encouraged and enabled by entrepreneurship blending innovation and business skills. This will provide the foundation for the next-generation of factories as products, having digital, adaptive and real-time features, and being able to create future generations of products.

5.2.2 Proactive Initiatives for Adaptive Manufacturing

Adaptive manufacturing is knowledge and intelligence-based and operates with state-of-the-art manufacturing and information and communication technologies and socio-technical systems. The adaptive manufacturing innovative models, methodologies and enabling technologies presented in this chapter support the manufacturing enterprises to face commercial challenges by promoting new and innovative paradigms clustered under several main groups. The implementation of adaptive factories advances towards a new and enhanced type of manufacturing systems which have to be more flexible – at a high performance level – responsive to the turbulent and permanently

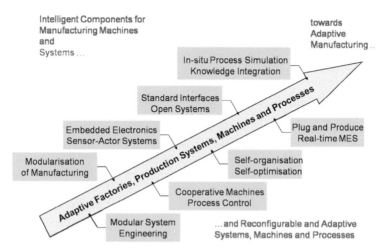

Fig. 5.7. ManuFuture Trans-sectorial Roadmap: Adaptive Manufacturing [5.8]

changing tasks and products (customised manufacturing) through the development of self-learning, self-optimising and co-operative control systems.

Adaptive manufacturing includes the reconfigurability of manufacturing systems and the implementation of knowledge in the operation system. "Adaptive production systems, machines and processes" aims at the development of adaptive technical components with high integration of electronics (mechatronic) modules, the implementation of the reconfigurability of the machines and the use of smart technologies for the manufacturing of plug-and-play components, employed in high-precision manufacturing.

The actions for embedding intelligence for enhanced processes aim at the development of cost-efficient monitoring systems which improve the prognostic capabilities, the reliability and performance of the monitoring systems. Adaptive tools and components as main entities of the adaptive manufacturing systems have a main contribution in the field by the in-situ process simulation, used to identify the behaviour of systems under usage constraints, and self-optimising drives and innovative electric-fluid energy sources. All these innovative and enabling adaptive technologies are in the present chapter shortly presented and graphically represented under the corresponding **ManuFuture Trans-sectorial Roadmap: Adaptive Manufacture** in figure 5.7. Their employment into manufacturing enterprises at all levels of abstractions for the implementation of the adaptive manufacturing is viewed as follows:

Short term with high priority, modular systems aimed at the development of the modularisation of manufacturing systems advancing towards a new generation of scalable and interoperable control systems, able to cope with the changing market demands;

Medium term with high priority, the enabling technologies grouped here aspire at implementing responsive factories through co-operative, self-organised and self-optimised behaviour of process control systems, and through embedded electronics and sensor-actuators systems;

Long term, future adaptive factories, production systems, machines and processes have to include as main components the plug-and-play elements required for high-precision manufacturing and the embedding of the new knowledge, and allowing integrating heterogeneous in-situ simulation models of manufacturing processes.

5.2.3 Proactive Initiatives Networking in Manufacturing

Due to the need to both conquer new and maintain existing markets, the number of manufactured product variants is rising steadily and product life cycles are becoming shorter. Also while trying to achieve cost optimisation, companies aim to creating lean processes with low inventories. As a consequence companies in a supply chain network find themselves in a difficult situation.

On one hand, they have to react to short-term changes of market demand and other events in the network and on the other hand long-term planning and co-ordination of the network has to be assured. This situation is further aggravated as changes in structural conditions such as the network topology or the selection of the network partners have to be adapted to in shorter and shorter time intervals.

Manufacturing units are embedded in regional and global structures. The focus of the past was the logistic chain. Networking of the future takes into account the relations to internal and external actors, influencing the dynamics and efficiency of all processes in the different chains. Synergy effects can be reached by concentration on the value chains and implementation of real time management, based on ICT. Regional relations and regional cluster play a dominant role in manufacturing.

Tomorrow's manufacturing units will work in complex, integrated and dynamic networks, often operating across the borders of companies and countries as each seeks to maximise their shares within the value chain.

As the scope and the dynamics of these production networks will increase continuously over the coming years, research and development has to tackle several areas in

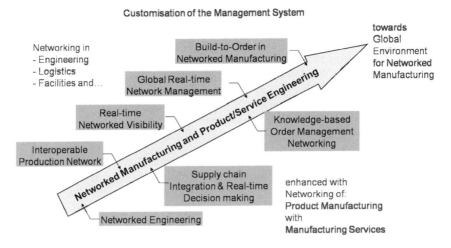

Fig. 5.8. ManuFuture Trans-sectorial Roadmap: Networking in Manufacturing [5.9]

order to come up with solutions regarding the network integration, the standardisation of interfaces and ICT systems, and enabling real-time decision capabilities throughout the network.

Looking in more detail at the production network, four different segments of the network can be identified:

- the customer and user network, including all organisations and processes bringing a final product to the customer or end user,
- the product and supplier network, including all manufacturing and service companies creating and delivering parts, components or raw materials and related services for the final product,
- the product engineering network, representing all activities across several companies to design and engineer a new or changed final product,
- the manufacturing system supplier's networks, involving all companies and processes for producing, installing and maintaining the production equipment, used for the manufacturing of products.

With the target of keeping or regaining a competitive advantage for European manufacturing, including the consequences of shortening product life cycles, requirements for faster reactions to changing customer demands and specific products, an integrative view of these four network segments is essential to achieving overall networked production.

Several enabling technologies are relevant and required for the implementation of this integrative approach to networked manufacturing have been identified and represented according to scale time and implementation priority in the following **ManuFuture Trans-sectorial Roadmap: Networking in Manufacturing.** The sequence of the technologies deployment is envisioned as in figure 5.8.

Short term with high priority, innovative strategies for networking manufacturing, new and innovative methodologies and technologies aiming at improving networked engineering and the interoperability of a production enterprise interlinked in a production network;

Medium term with high priority the real-time logistics network has to be investigated and new management models recommended for global and real-time manufacturing networks, network visibility has to be implemented and the supply chain integrated to allow real-time decision making in non-hierarchical networks;

Long term, the global environment for networked manufacturing ultimately aims at implementing knowledge-based and adaptive manufacturing through intelligent order management and factories and logistics on-demand concepts.

5.2.4 Proactive Initiatives for Digital, Knowledge-Based Engineering

Manufacturing Engineering is a holistic approach which includes the engineering of product design engineering, the factory structure, the development of the organisation and operational management system, process engineering and the set-up of the required resources (buildings, infrastructure and media supply, machines, tools, equipment, processes, logistics, monitoring systems, staff). A critical part of manufacturing engineering is the ramp-up phase of new products.

At all levels, e.g. manufacturing network, segment or system, machine or equipment, subsystems and processes, the factory and its manufacturing processes can be defined in their current and/or future states. Planning and engineering need the knowledge of process technologies as well as experience from the past. The objective that a factory operates permanently at the optimal economic point and with high reliability, flexibility and productivity, requires active tools and methods for optimisation.

It is necessary to develop and implement integrated tools for manufacturing engineering with a core in standardised factory data management and a full set of specific tools from process planning to multi scale modelling and simulation. There is also the opportunity to realise a European platform for manufacturing engineering.

As knowledge represents the main source for innovation and implementation of digital and virtual factories and products, the research and application field mentioned can be structured under the cluster of enabling technologies and tools "digital and knowledge-based engineering". As major objective of the European industrial sector is to continue to play a leading role at global level, factories have to be approached as a new and complex type of product.

They are long-life products which have to be continuously adapted to the needs and requirements of markets and economic efficiency. Many of the influencing factors are continuously changing and sometimes these changes are turbulent and unpredictable.

Factories operate in networks and are parts of different process chains: the design and engineering of the products, supply chains from customer orders to customer delivery, supply chains for consumable materials and waste, supply chains for factory machines, equipment and tools. Manufacturing engineering is a multi-dimensional discipline with the key role in factory operations.

The innovative and enabling technologies and tools identified as crucial for this proactive field are grouped according to sustainable digital factories and products: design, modelling and prototyping; virtual factory simulation and operation; real-time

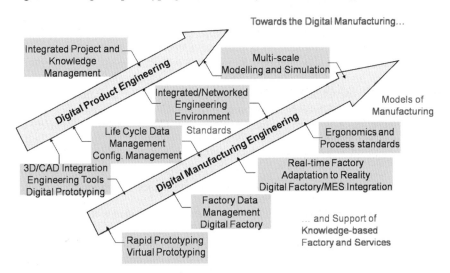

Fig. 5.9. ManuFuture Trans-sectorial Roadmap: Digital, Knowledge-based Engineering [5.10]

(smart) factory, and process modeling, simulation and management and they are graphically represented in the **ManuFuture Trans-sectorial Roadmap: Digital, Knowledge-based Engineering** (figure 5.9).

These technologies are planned to be deployed in time scale and priority as follows:

Short term with high priority, the collaboration of digital manufacturing engineering with digital product engineering through rapid prototyping of the virtual factory synchronised with the 3D/CAD integration of engineering tools and digital prototyping of virtual products;

Medium term with high priority, factory data management for 'Life Cycle Data Management for the Digital and Virtual Factory'. The development of the virtual factory framework aiming at integrating heterogeneous and autonomous technologies and tools for planning, design, manufacture and implementing the two entities in digital and virtual states. The real-time factory represents one of the desired goals, integrating and synchronising the digital factory with real-time data to allow the data model to adapt to and reflect reality.

Long term, multi-scale process modelling, simulation and management to implement the holistic approach to manufacturing engineering at all its scales, from network to manufacturing processes and states, from digital, virtual to real-time. Scientific research can significantly contribute by describing all process phenomena and consequently process models. Thus, zero-defect manufacturing, high capability and high performance can be reached using science-based simulation.

5.2.5 Proactive Initiatives for Emerging Technologies

Fast, reliable, capable, low cost, and flexible are the major criteria for the development of manufacturing technologies. The main objective of this field of initiatives is to overcome existing product or process limitations by developing innovative solutions for new and emergent products and new and emergent technologies.

Processes should be high performance, capable of achieving at micro and nano scales in dimensions and precision, low resource consumption, allow rapid prototyping and manufacturing, of light-weight and with function integration. This requires the continuous development of processes and process chains. This is a field requiring significant co-operation between material technologies (i.e. graded and adaptive materials, surface technologies) and manufacturing processes.

The implementation and development of "next-generation European manufacturing technologies" is enabled by a set of "emerging technologies". These are defined as the manufacturing and engineering technologies and tools whose scientific basic principles and theory are understood and well formalised and where at least some useful and widely accepted fields of applications and studies are recognised. Despite identified and partially proved applicability, their potential is mostly unfulfilled as evidenced by a lack of significant products and possibly by the lack of market demand or need.

They are of high potential, as the risk taken by the pioneers in implementing them is ultimately expected to be recompensed by competitive advantage, adaptability and high flexibility in order to cope with the market demands and the changes and challenges driven by new products and their related services.

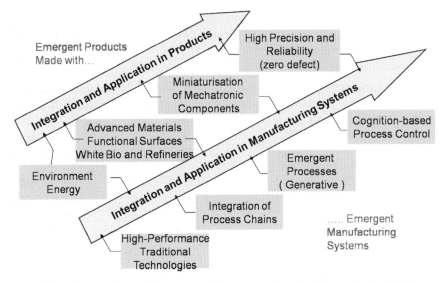

Fig. 5.10. ManuFuture Trans-sectorial Roadmap: Emerging Technologies [5.11]

Several emerging technologies relevant for the next-generation European manufacturing systems have been identified and structured according to the following four main clusters: environmental and energy technologies, performance and efficiency-oriented technologies, advanced material engineering and product-oriented technologies.

Distributed on the time scale and priority, these have been graphically represented in the **ManuFuture Trans-sectorial Roadmap: Emerging Technologies** (figure 5.10) with the following deployment by synchronisation, collaboration and harmonisation of integration and application in products and manufacturing systems:

Short term with high priority, the combination of the high- performance traditional technologies, when pushed beyond their limits and capabilities, with low energy consumption and scavenging technologies to allow the implementation of clean manufacturing processes.

Medium term with high priority, fully integrating advanced materials and materials engineering, functional surfaces and bio-processing into the process chains for the purpose; the further miniaturisation of mechatronics components.

Long term, Emergent and generative processes, cognition-based process control of high-precision and reliable (zero-defect) manufacturing processes.

5.2.6 Proactive Initiatives ICT for Manufacturing

The cluster represents the collaboration and harmonisation of state-of-the-art technologies from both ICT and manufacturing engineering technologies. It is anticipated that the exploitation of emerging technologies in each field will be enhanced by applications in the other.

Drivers of information technologies are electronics and information systems. Many innovations made for home or business can be transferred to the manufacturing sector.

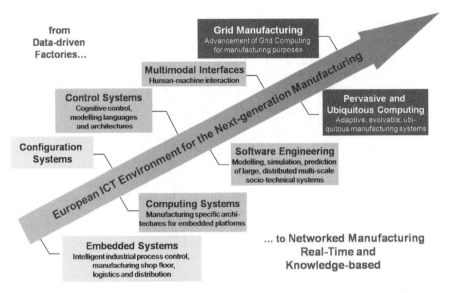

Fig. 5.11. Trans-sectorial ManuFuture Roadmap: ICT for Manufacturing [5.12]

However, there are specific requirements driven by manufacturing sectors which need specific developments. The main requirements are:

- open systems and architectures of IT systems for manufacturing,
- global platforms and standard interfaces for distributed engineering and network management,
- a global platform for life cycle management of technical products,
- vision technologies,
- efficient interaction of humans and machines,
- solutions for technical intelligence, process monitoring, diagnostics, navigation,
- real-time operations from process to networking ,
- knowledge management based on data collection, analysis of history and experiences in the life cycle of products,
- reliability of systems in high efficient manufacturing execution,
- grid technologies for manufacturing engineering.

Excellence in the integration of ICT and manufacturing can be conducive to the identification of new applications and as yet un-fulfilled potential.

An instance of this opportunity is that given by the increased capabilities required for the next-generation of European manufacturing systems such as their adaptability and flexibility, and their digital and knowledge-based approach. Manufacturing industries are one of the strongest markets for ICT. ICT for manufacturing enabling technologies and necessary tools are graphically grouped in the **Trans-sectorial ManuFuture Roadmap: ICT for Manufacturing** (figure 5.11).

These technologies are planned to be implemented in the manufacturing enterprises as follows:

Short term with high priority, configuration systems aiming at production/services customisation will enable manufacturing enterprises to meet the customers' individual requirements more effectively, by providing more customisation approaches related to the design and development of the new generation of modern products and manufacturing systems.

Medium term with high priority, the development of computing and embedded technology for the digital factory as a generic platform which embeds all state-of-the-art modelling, simulation, optimisation, and visualisation technologies and tools for turning the digital and virtual factory into reality. Digital libraries and content for engineering and manufacturing and cognitive control systems: modelling technologies and architectures. The development of an integrated framework for networked multi-modal collaboration in manufacturing environments aimed at enhancing the interfaces human-machine-human through new and innovative, easy and friendly modes of interaction.

Long term, grid manufacturing aimed at migrating the grid computing technologies and tools for coping with the challenges of networked manufacturing, particularly the requirement for high-flexibility. Pervasive and ubiquitous computing is enabling and emerging ICT to support the implementation of the adaptive evolvable ubiquitous manufacturing systems.

5.3 Proactive Initiatives for Factories of the Future

Factories are extremely complex products which follow product innovations and market requirements. They have a long life time but they have to be adapted to the state-of-the-art and value stream continuously. No other products have comparable characteristics. Systems efficiency depends on the properties of its elements, their relations and the structure of the system. That is why some technologies, defined in the Roadmaps have high potential for the efficiency of elements and are enablers, but they have to be integrated in the factories by balancing and optimising the holistic system factory. Experts estimate the potential of productivity by implementation of some enabling technologies and strict focus and concentration on the value stream at about 50% and more.

As an example, an investigation of the cost savings of new technologies in a factory five years ahead shows:

- 40% for the implementation of high performance technologies,
- 30 % for the implementation of next generation of machines,
- 25% without adaptation of the system,
- 40% with adaptation and balancing of the system.

The conclusion for the strategic development towards factories of the future is clearly the integration of innovative solutions, and system orientation is the key for the efficiency of factories. Figure 5.12 illustrates the generic model for system-orientation of factories. The factory is understood as a complex network in dislocated plants in which OEM and suppliers co-operate in the value chain from basic materials to delivery to usage including the services and recycling (end-of-life).

This high level system has process chains (value streams) and can be divided in subsystems along the processes in which the value is created. Criteria of optimisation are cost, time and quality of all processes and the efficiency of the systems. Following the CSD paradigm, social and ecological criteria have to be added.

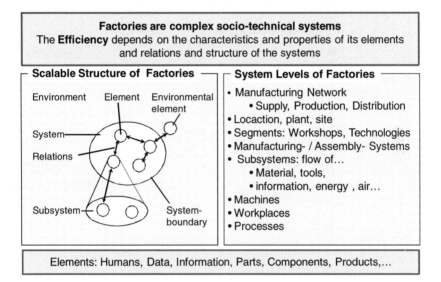

Fig. 5.12. Holistic View on Factories: Elements, Relations and Subsystems (© IFF/IPA) [5.13]

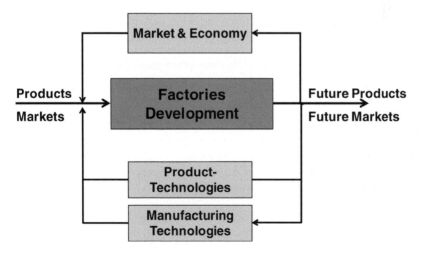

Fig. 5.13. Usual Factories Strategic Development (© IFF/IPA) [5.14]

The development of factories (figure 5.13) usually follows the requirements of products, markets and adaptation of resources in order to activate short-term economic potential. Nearly all influencing factors have a dynamic behaviour so that the environment around factories is turbulent. The factory life time extends over the life time of several product generations.

Customisation and the shorter life time of products make it necessary to continuously adapt the 'factory' system. Adaptation of the system and sub-systems causes inefficiencies when the systems are not open and changeable in an effective way.

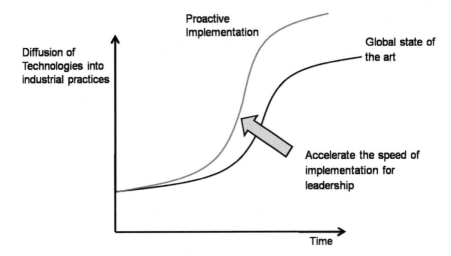

Fig. 5.14. Fasten the Speed of Implementation for Leadership (© IFF/IPA) [5.15]

The flexibility and adaptability depends on the technical and organisational capacity and capability of the installed resource including buildings and machines, and on subsystems like logistics and the competences of the humans from operator level to management (skill levels).

Nearly 200 organisations such as supplier, services, consultants, experts, engineering companies, producers of equipment, educational organisations, social organisations, unions, local and states governments as well as research institutes contribute to the process of adaptation and can influence the efficiency of the 'factory' system.

They all need to be orientated to the future vision and strategic development of factories in general as well as to the concrete plans for the specific changes to the particular factory system. This makes it necessary to implement factories of the future and secure and share experiences of the structural changes required to transition from the factory system of the past to more competitive and sustainable factories. Enterprises consequently have to change internal procedures for factory development from reaction to short-term events to long term strategic development. The character of factories and their structure usually changes when the products and market requirements change or when the economic pressure increases. Companies usually implement available and reliable technologies or methodologies for the development of factories. They can take advantages by implementation of the state-of-the-art but cannot take leadership unless they innovate.

The ManuFuture road is characterised by proactive and fast implementation of innovative solutions developed by co-operative research and following the CSD paradigm. Leadership in technologies offers new markets or expanding market shares in conventional and emergent areas of global manufacturing for manufacturers of factory equipment and its users. New solutions will change the structure of factories including the role of workers and the relationships between suppliers and customers. The competitive position of European factories depends on the speed of implementation of product and manufacturing technologies – proactive initiatives – and the market development or potential. Rapid and early implementation of the actions defined in the Roadmaps of ManuFuture increases exposure to technical and economic risks. Also the cost benefit of technology development is lower than that achieved by quickly following the trends.

However, it is known, that companies with aggressive innovation define the standards and control technical development. They have advantages in the global market against fast followers when they are able to place innovations in global market place on time. This makes it necessary to develop technologies in close co-operation with (leading) customers and to gain practical experience as fast as possible (see figure 5.14).

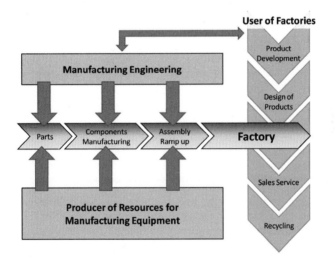

Fig. 5.15. Synergetic Development of Factories (© IFF/IPA) [5.16]

All of the actions for research defined above require close co-operation between producers of factories systems, based on shared strategies for the production system of the future. The producers of factories play a key role as an enabler for the strategic development of factories in cooperation with their customers (figure 5.15). The effective and timely implementation of the Roadmaps and consequent economic results in world markets are only constrained by the requirements for developing collaborations between the actors. The Roadmaps will change the structure of the manufacturing

system we call the factory and push European manufacturing towards new generations of factories including the knowledge-driven factory of the future. More than one
road of the trans-sectorial actions will take European manufacturers to leading positions in the global market. Some are trend-setters – some have a high economic potential. The vision of ManuFuture is the implementation of a knowledge-driven factory
of the future to which all of the 'Roads' contribute, each in a more or less significant
way. Implementation in practice will depend on the characteristics of each business
area and its strategic orientation. Business specific objectives and many influencing
factors (markets, competitors, financial, political, technical, regional, social etc.) will
constrain structural development.

However, all will need fast implementation and the requirement to share the experiences of the first applications of these new approaches to develop the solution
embodied in their individual manufacturing system. Manufacturing and technology
companies are the enablers of the structural change from state-of-the-art to the Factories of the Future (figures 5.16 and 5.17).

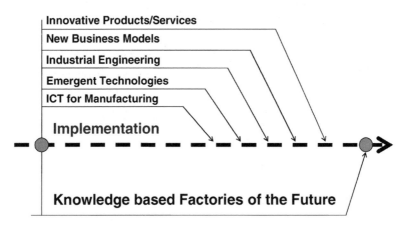

Fig. 5.16. Implementation of the Proactive Action in the Factories of the Future (© IFF/IPA)
[5.17]

Many of the companies in this sector are SMEs and many of them do not have the
financial resources to implement high risk innovative manufacturing technologies or
to change their structure in a medium-term time frame.

This makes it necessary to implement light house projects in co-operation and realise highly developed factories as pilots with the support of public funding. To increase the speed of diffusion into practice it is necessary to support the mechanism of
experience and knowledge transfer via an efficient infrastructure and European networks. In the following sections example light house projects, in which the primary
and ambitious actions of the long-term trans-sectorial Roadmaps can be realised, are
introduced.

They follow the CSD paradigm and are specifically orientated to the knowledge-
based factories of the future. Each pilot has high economic and technical potential and
they are called (figure 5.17):

- Transformable Factories,
- Networking Factories of Excellence,
- Learning Factories,
- Factories of Emotions.

This selection is based the potential for leadership and rate of innovation in:

- customisation and customer-oriented manufacturing,
- activation of the technical potential by high performance,
- increasing the efficiency of all resources (sustainability),
- activation of the human potential and skill including knowledge.

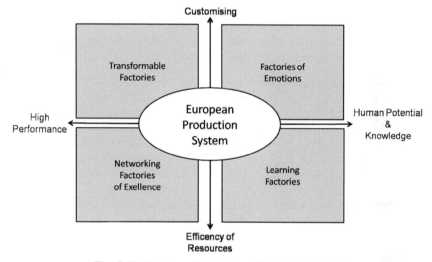

Fig. 5.17. Factories of the Future (© IFF/IPA) [5.18]

They all take into account that factories are complex systems and need permanent adaptation to the requirements of product and process innovation to realise economic benefit. Each of the proposed pilots follows visions from the ManuFuture Roadmaps.

5.3.1 European Production System

Production systems approaches form the fundamentals that guide factory development. Examples such as the Toyota Production System have shown their potential impact on the long-term competitiveness of manufacturing. Methodologies for management and optimisation of technical systems – influenced by the philosophy of lean thinking – are today's standards in all areas of production. They define the parameters of productivity and economic efficiency. It is essential for Europe to define a long-term holistic production system to reach its objectives for competitiveness and sustainability. The European production system has to define the standards for manufacturing: economic, ecological, social and has to recognise European cultural traditions as well as the promise and possibilities of innovative technologies. There are possibilities that form a starting point for this work which are mainly based on the knowledge of skilled people and the organisation of work. The ManuFuture Road-maps contribute by including technological targets to exceed the state-of-the-art, rapid engineering, customisation and new business models. Europe now has the chance to lead the production systems of the world.

5.3.2 Transformable Factories

The main objectives of past developments in manufacturing were characterised by high efficiency of resources (productivity) and rapid adaptation of manufacturing systems to the task (flexibility).

A major trend for factory development in the recent past was to leverage flexible automation and Computer Integrated Manufacturing (CIM) to activate the potential of productivity and flexibility in the current operation.

Today, again technologies offer a new potential for high performance and flexible factories particularly for *mass customised* production which leads to:

- high grades of flexible automation (robots, CNC machines) characterise the technical concept of high volume and low cost factories. Highly sophisticated technical solutions operate extremely quickly or with high utilisation and reliability. Operations and sequences can be changed with low set-up costs. Tools and equipment are specialised for ideal operations. The number of direct workers required decreases to a level where the costs of labour are minimal. Regional conditions of work lose their influence on the manufacturing costs. Utilisation depends on the quality of service.
- High precision factories operate beyond the state-of-the-art in accuracy and quality. Significant technology areas are in the sectors of tools, moulds and dies used for forming, casting or non-metallic parts manufacturing. These are suppliers for nearly all sectors of manufacturing. High precision requires highly skilled workers and capable processes. Factories for high precision have to be equipped with measurement tools and systems for monitoring and control of processes. These types of factories are familiar with tolerance requirements in sub-µm dimensions. Many of them need an extreme infrastructure for manufacturing (climate, air, fluids, protection against contamination)

Transformable factories include the aspects of both high performance and high precision. In addition they must react quickly and without loss (lean) to changing markets and changing products. Solutions historically used universally applicable machines and operated with standard processes in the organisation.

New generations of manufacturing systems are able to change the structure of manufacturing by exploiting open supply-networks, flexible use of competences and organisational change. Fast reconfiguration of manufacturing and assembly systems is one element of transformability.

The structure of the factory is characterised by a system of self-optimising, self-controlling and flexible operations management. Transformable factories have new generations of ICT to manage change at all levels and along process chains, such as: smart factories, RFID, online simulation, open information and communication systems. The main objective of the transformable factory is to always run on the best economic operating point by the permanent and continuous adaptation of internal and external processes.

5.3.3 Networking Factories of Excellence

Experts estimate that the economic potential of improving networking in manufacturing is about 25 % of the total costs. The automotive industries and other series manufacturers have made significant progress in the efficiency of supply networks. Examples show the advantages of the harmonisation of logistic chains as for example in the use of supplier parks or along supply chains.

The efficiency of networking in manufacturing depends on the distance between operating actors. Short distances have the potential for the reduction of time and cost, when changes of products or capacities – influenced by dynamic markets – are required. The standardisation of interfaces is necessary. These approaches are not only a strategic option for automotive industries but are applicable to single and small series production.

Small and medium enterprises mainly focus on products with high complexity, low lot size and high number of variants. Specialisation and market niches characterise the development of products. Efficient networks operate without hierarchies and are based on the principles of reliable co-operation and work sharing.

Networking is essential in engineering and technology as well as in the sales or service areas. There are many medium sized companies with more than one manufacturing location and with local foot prints. It is evident, that regional networking and co-operation in the chains from engineering to products or from parts manufacturing to assembly and delivery has a high innovation potential. The speed of innovation depends on the synergy given by regional networking and the necessary harmonisation of interfaces.

At the core of networking factories of excellence are the methodologies required to effectively manage the networks and to operate in real-time environments. A pilot initiative for regional networking is of fundamental interest for the structural development of regions in Europe.

5.3.4 Learning Factories

Organisational learning is a well-known methodology for industry. It leverages individual learning (skill, experiences), learning as methodology for permanent improvement of processes or leaning by benchmarking. Learning organisations have been implemented in many enterprises as a part of their management system.

Learning is a process to reduce losses or improve productivity by analysis of past operations, usage of experience, creating new solutions and activation of explicit or implicit knowledge. Learning factories are characterised by the implementation of learning systems in technical solutions, self-adaptation, cognitive IT, proactive improvement and integration in the holistic production system.

To activate the full future potential of learning factories, some new areas are of significance for all manufacturing sectors:

- implementation of scientific based process models in the control and monitoring systems,
- history analysis and data management,

- knowledge-based engineering,
- digital factory – connected with the real factory (virtu-real)
- integrated system support for optimal solutions,
- integration of external knowledge,
- training and skill.

The pilot initiative of a learning factory is one approach to exceeding the state-of-the art in production systems and technical intelligence in manufacturing.

5.3.5 Factories of Emotions

The relationship of manufacturers to their customers is one of the critical factors for a business. Many companies see market and customer-orientation as a part of the overall business culture and have consequently changed their management, methodologies, structure and systems to a high customer-orientation. Standards of quality management, order management, the logistics system and services focus on the customer. Automotive industries have developed solutions to reduce the time from customers request to delivery of products to a minimum.

The "Gläserne Fabrik" (transparent factory) for VW cars in Dresden invites customers to watch the assembly of their product. Customisation reduces the lot size for manufacturing, increases the variants and requirements for special solutions and costs of services in comparison to mass production of standardised products. Following this customer orientation and our understanding of a factory as a complex socio-technical system, it seems possible to develop and realise next generations of" factories of emotions":

Factories are the place where people work, owners and investors make profit, customers products are made, suppliers deliver parts and components, public organisations and governments influence the processes, collect data supervise, etc. Many people and many organisations – inside and outside – influence the operations and the efficiency of manufacturing. They are direct or indirect interactive partners of enterprises. The main question is how to integrate these players to best benefit.

The vision of factories of emotions is a system inspiring positive emotions to products, processes and management by the implementation of a production system with:

- high transparency and high reliability, high efficiency,
- short distances to the locations where the customers (and markets) are, both internal and external (design to workplaces to customers),
- efficient and non-bureaucratic internal and external communication (new solutions of ICT),
- human-centered organisation and management,
- human-oriented workflow, flexibility in capacities, high skill, high social competences,
- innovative human-machine interfaces, ergonomic work places, etc.
- lean and zero defects, workshop-oriented layout,
- life cycle management,
- sustainability integrated in company strategy and implemented, methodologies.

The factories of emotions follow the ManuFuture CSD paradigm in management and layout. The model integrates many actions of the Roadmaps, such as ICT for manufacturing, and focuses on the interaction of humans with the manufacturing area internally and externally. It reflects on the fact, that about 200 external organisations have different interests on what happens inside and in the periphery of factories.

It gives the possibility of implementing significant innovations in the business model and to activate external competences to the benefit and development of factories.

References

[5.1]   Paci, A.M.: Transectoral Technology Impact on Sector - EPPLab - Emerging Production Paradigms Laboratory – Annex to Leadership project "Overall Manufuture Roadmap" (June 15, 2007)

[5.2–5.5]   Westkämper, E.: Private Archive

[5.6–5.12]   ManuFuture Conference Tampere (Fin) (2006)

[5.13–5.18]   Westkämper, E.: Private Archive

6
The ManuFuture Road to High-Adding-Value Competitive Sustainable Manufacturing

F. Jovane and E. Westkämper

The ManuFuture road to High-Adding-Value Competitive Sustainable Manufacturing, as emerging from the results achieved and the foreseeable perspectives is outlined.

ManuFuture, acting as a strategic infrastructure to pursue CSM, has generated SI and the related implementation framework. This encompasses from reference models for action and global co-operation, to EMIRA, to the 25 national ManuFuture platforms and KIC. Stakeholders, from public authorities and financial institutions, industry, university, research institutes and centres, are co-operating in SI implementation, through basic activities. To speed up and lead the implementation process, pilot initiatives are being explored and developed. A manufacturing joint Technology Initiative (JTI) is currently being considered. Its objective is the implementation of manufacturing enabling technologies of the future, based on the ManuFuture technology Strategic Intelligence (SI).

The Eureka cluster ManuFuture industry has been conceived and is being launched. It will implement the ManuFuture SRA vision concerning European production systems: products for the world market and processes to retain production in Europe. More than 40 companies, supported by ten research institutes, making up a KIC, will be investing in the cluster. Overall value of the projects: 400 MEURO.

6.1 Towards European HAV K-Based CSM: The Current and Perspective Role of ManuFuture

As presented in the previous chapters, Europe should play a leading role in the technological and industrial global revolution that should address competitive sustainable manufacturing. ManuFuture must contribute to such a revolution through the generation and implementation of:

- Strategic Intelligence (SI): i.e. the ManuFuture Vision 2020, the Strategic Research Agenda (SRA), Roadmaps; this is necessary to guide the aforementioned transformation processes,
- the framework with implemented SI, to pursue CSM.

The ManuFuture Platform may be considered as an infrastructure, acting within the innovation cycles (see figure 3.2). It has set up a rolling process to manage the:

- SI life cycle: i.e. from foresight, to roadmapping, implementation, monitoring,
- ManuFuture framework life cycle: i.e. from definition, to implement-tation, management, monitoring,

- management of implementation of SI within ManuFuture framework to pursue CSM.

The ManuFuture platform has gone through the definition phase, generating the SI presented in chapter 4 and 5 and the main features of the ManuFuture framework. The ManuFuture platform is now in the implementation phase. It is setting up the framework, managing and further developing it. Herein SI is being diffused, adopted and used by PAs at EU, national and regional level, as well as the IK-T, to pursue CSM. These may be considered as part of ManuFuture basic activities. Pilot initiatives are in progress to focus and accelerate basic activities. All this is reported in the following paragraphs.

6.1.1   The ManuFuture SI implementation Framework

The main features of the ManuFuture SI implementation framework may be summarised as follows:

- the reference model for basic activities and pilot actions at European level, figure 6.1. It was presented at the ManuFuture Conference 2007, held in Porto, where convergence of stakeholder thinking to CSM was assisted by a polar star analogy where the polar star is a beacon giving direction and radiating SI,
- the reference model for co-operation at global level in co-operation with CIRP including the basic process for governance: generation, diffusion, adoption and use of SI: i.e. Vision, SRAs, Roadmaps, for transformation of both, industry and E&RTD&I system , as a part of the SI life cycle,
- the EMIRA environment as the multi-level space for manufacturing innovation and research within ERA, where the E&RTD&I system operates to sustain HAV K-based industry action on the market,
- the stakeholders and the manufacturing KIC,
- MF-EU and 25 national/regional platforms, for managing basic actions pilot initiatives. Monitoring for governance.

They may evolve throughout the ManuFuture framework life cycle.

6.1.2   ManuFuture Reference Model for Basic Activities and Pilot Actions

The system, based on E&RTD&I Intelligent K-Triangles (IK-T), must act within EMIRA, as the generator of High-Adding-Value. This must be embedded in products/services, processes and business models, to enable and sustain a HAV, K-based, CS European industry.

The complex interaction between stakeholders, their activities, domains of action, policies, strategies, and so forth, may be represented as shown in figure 6.1. It concerns from basic to pilot initiatives, the latter being aimed at leading and accelerating the pursue of CSM.

This is the "Steering EMIRA" reference model for basic activities and pilot actions, developed by ManuFuture to increase effectiveness and governance of the E&RTD&I system for manufacturing, while aiming at developing and implementing

CSM in Europe, in co-operation with other countries and global regions. The reference model as shown in figure 6.1, relates:

- the domains, market and EMIRA,
- the role of public authorities, impacting at macro level on industry and academia, following the triple helix approach,
- the European E&RTD&I system, implementing education research and innovation at local level,
- the strategic leading role of the evolving polar star, made up of the SI (Vision, SRA and Roadmaps) developed by ETPs and the like,
- the leading effect of the evolving polar star on the actions of all stakeholders: i.e. PAs, at macro level, industry, university, research institutes, making up the intelligent K-Triangles, at local level,
- the consequent convergence of PAs programmes and initiatives and E&RTD&I (IK-T) projects (local level) towards the common strategic goal: i.e. CSM, while governing its development, implementation and upgrading process.

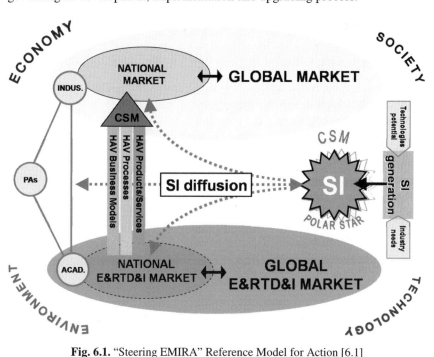

Fig. 6.1. "Steering EMIRA" Reference Model for Action [6.1]

Following the above reference model, results already obtained and in progress are reported in the next paragraphs with future actions.

6.1.3 EMIRA

As described in chapter 3, the European Manufacturing Innovation and Research Area (EMIRA) – introduced by ManuFuture SRA – is a multi-level domain where E&RTD&I activities

- are carried out by, university, research institutes and industry – acting within the knowledge triangle [6.2, 6.3], (see figure 6.2) – in co-operation with public and private institutions : i.e. from PAs to financial institutions,
- are complying with strategies that, – depending on the level concerned – may come from technological platforms (SI: i.e. Vision, SRA, Roadmaps) as well as company business plans.

To contribute to the establishment of the new CSM and to sustain its development, in the evolving ESET context, EMIRA must be:

- internally efficient and effective in implementing the K-generation, diffusion, use process, through the education, research, innovation I-K-Triangle,
- generating and embedding into products/services, processes and business models, the knowledge that will gain them high value recognition by a competitive and sustainable global market,
- globally competitive and sustainable.

EMIRA is a component of the ERA, dedicated to manufacturing. Following the green paper "The European Research Area: New Perspectives" and a ManuFuture survey, EMIRA should, mainly:

- be a sharing knowledge area (notably between research and industry),
- develop world-class research infrastructures,
- optimise research programmes and priorities.

6.1.4 ManuFuture National and Regional Platforms

To be representative, to involve a vast number of European companies, sectors and regions and to have a broad impact, the ManuFuture platform promoted the creation of platforms at national and regional levels (NRTP). From the first initiatives (dated from 2005, when Spain, Portugal and Italy officially launched their platforms), there are today 25 NRTP, in different stages of development, involving hundreds of organisations and thousands of people. The NRTP represent an army for the implementation of the ManuFuture strategy, since they can have a critical role in ensuring the European co-verage (at national and regional levels) and form the critical mass for such an initiative.

They are mobilising partners, particularly from industry and especially SME's, facilitating dissemination and demonstration activities much closer to the target audiences, promoting and managing complementary projects and initiatives, at national and regional levels (stand alone or organised in European networks (like EUREKA or ERANET), mobilising private and public funds, at national and regional levels, mobilising national and regional authorities and public bodies and influencing the creation of national and regional funding programmes, complementary and, preferably, aligned with the European programmes.

They are interlinked with the ManuFuture European platform, thus helping to structure the multi-level domain, covering from EU to national, regional and local level: i.e. EMIRA.

Manu*future* National Platforms and Initiatives

- **Registered 25 MF NTP**
- 12 for EU15
- 12 New Member States
- 1 Switzerland

Fig. 6.2. ManuFuture National Platforms and Initiatives

The main features and activities of the ongoing ManuFuture national platforms and initiatives are reported in the catalogue of ManuFuture national and regional technological platforms [6.4].

6.1.5 Manufacturing KIC

Within the European Manufacturing Innovation and Research Area (EMIRA) stakeholders are contributing to CSM. They encompass from university to research institutes and centres, technological and sectorial ETPs, excellence centres, networks of excellence, integrated projects, industrial organisations and associations, public/private research organisations for application oriented research. They interact with EU FPs, Eureka, national and regional E&RTD&I programmes and initiatives.

As envisioned by the ManuFuture Vision 2020, they make up the Manufacturing Knowledge and Innovation Community (KIC) [6.5]: the largest and strongest in Europe and, at this moment, in the world!

All stakeholders are involved – through various manufacturing related Technological Platforms, TPs – in the SI generation, diffusion, adoption and use process.

6.1.6 The SI Generation- Diffusion- Adoption and Use, Monitoring Process

The generation, diffusion, adoption and use process and its monitoring, taking place within the current ManuFuture framework [6.6] are shown in figure 6.3. They are enabled and supported by basic activities and pilot actions. Manufacturing KIC stakeholders [6.5] – ranging from platforms to individual researchers – are involved in the SI generation, diffusion, adoption and use process.

- Generation phase: technological platforms are meant to generate SI, by involving stakeholders, mainly industry, university and research institutes. PAs are close observers. As presented in chapter 3, 4 and 5, the ManuFuture European platform has extensively contributed to SI (Vision 2020, SRA and Roadmaps) generation. ManuFuture national platforms and initiatives have taken a similar approach. European/national platforms are contributing to make up a consistent and useful evolving polar star by providing visions, SRAs and Roadmaps.
- Diffusion phase: this phase involves all stakeholders from macro to local level. It is a costly and complex exercise. In particular the role of industrial associations, professional organisations as well as scientific and technological societies is fundamental for diffusion to stakeholders.
- Adoption and use phase: this phase plays a fundamental role for multiple coordinated as well as synergic actions to generate, implement and upgrade CSM. All stakeholders, from PAs to SMEs, may play an active role.

SI components, i.e Vision, SRA and Roadmaps; stakeholders, at manufacturing as well as EMIRA levels, are reported in table 6.1. This shows, in broad terms, how results may be used by single stakeholders for RTD&I activities or by organised subsystems, for strategic governance. Adoption and use is taking place, both at PAs and local level, but following different paths.

Table 6.1. SI Components, Stakeholders, EMIRA Levels, Governance Matrix

	RESULTS	PAs			INDUSTRY			UNIVERSITIES		RESEARCH INSTITUTES	
		EUROPEAN	NATIONAL	REGIONAL	EU ASSOC	NAT ASSOC	LOCAL	GLOBAL	FIELD	GLOBAL	FIELD
VISION	STRATEGY	✚	+	*+*	✚	*+*	+	✚	+	✚	+
SRA	ET: DIRECTIONS	*+*	✚	+	*+*	✚	*+*	*+*	*+*	*+*	*+*
ROAD-MAPS	RTD FIELDS TIMES	+	*+*	✚	+	*+*	✚	+	✚	+	✚

At macro level, PAs are investing resources using adopted SI. The European Framework Programme (FP7) has already adopted SI, generated by ManuFuture and other European technological and sectorial platforms. Such SI is being used for NMP work programmes and related annual calls. Some national as well as regional RTD&I programmes are acting similarly. The new Eureka strategy in progress will refer to SI as shown above. The impact of aforementioned activities is impacting on academia and industry, following the triple helix model [6.7].

At local level, ManuFuture SI has been and will be used by stakeholders: i.e. industry, university and research institutes and centres, (the intelligent K-Triangles) to carry out E&RTD&I activities addressing CSM. To generate the best possible research – market high-value chain, intelligent K-Triangles should have an embedded intelligent router (figure 6.4), that would refer at the evolving polar star and its SI content.

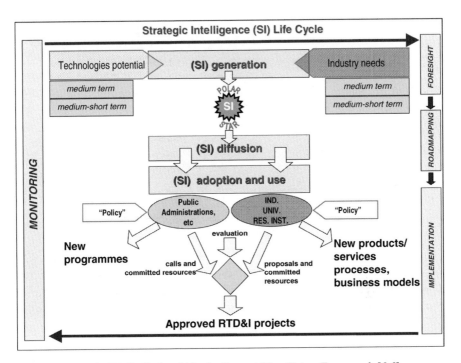

Fig. 6.3. SI Life Cycle within the Current ManuFuture Framework [6.6]

As IK-T stakeholders work on different time horizons and may be far from the SI language, a process is required to connect all stakeholders to the evolving polar star. Such a process could be implemented by responsible institutions, such as associations for industry.

It is worth underlining that European regions (see figure 6.5) are playing a fundamental role in E&RTD based innovation. The evolving polar star mechanism can be of help to them.

6.1.7 Basic Activities and Pilot Actions at Global Level

As stated before, it is necessary that co-operation among countries and regions, particularly on sustainability, take place to pursue CSM. Using the ManuFuture reference model for basic activities and pilot actions, a study has been carried out in co-operation with CIRP, within the 2008 CIRP KN Paper [6.9].

The study shows that PAs in advanced countries and emerging countries are fostering and supporting CSM, through E&RTD&I. At the same time they are investing in related international co-operation activities. It also shows the existence of a mechanism complying with the ManuFuture reference model for action in each country.

A form of evolving leading mechanism can be traced, that can be used by PAs and for IK-Ts activities. A global reference model is emerging from the CIRP study [6.9].

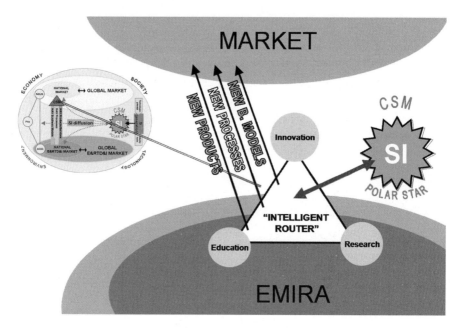

Fig. 6.4. Strategic Intelligence SI Adoption and Use at Local Level [6.1]

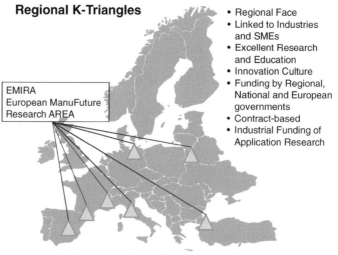

Fig. 6.5. Regional Knowledge Triangles (© IFF/IPA)[6.8]

It shows that the co-ordination of the various evolving polar stars and their content may help a synergic global use of strategic resources, as previously explained. Co-operation among countries and regions, particularly on sustainability, takes place. It would support, as described in paragraph 1.1, decisions by PAs and Intelligent

Knowledge Triangles (IK-Ts). Co-operation and competition must be carefully balanced.

6.2 ManuFuture Basic Activities and Pilot Actions

Because of the size and complexity of the transformation process, that concerns European industry and the related E&RTD&I system: i.e. the domain of action, ManuFuture activities addressing CSM are of two kinds:

- basic activities, to set and govern the Manufuture framework, to implement SI, thus igniting the transformation process,
- pilot actions, to lead and accelerate such transformation process.

6.2.1 ManuFuture Basic Activities

ManuFuture basic activities concern – on a rolling basis – from manage-ment of SI and FW life cycles, to the implementation of SI within FW, to monitoring and governance of the above processes, while interacting with the stakeholders, within the evolving ESET context.

Achieved results concerning SI and FW generation have been presented. SI diffusion, adoption and use concern various stakeholders. European, national and regional PAs are embedding ManuFuture SI within their E&RTD&I programs and FP7 uses it for its work programmes within NMP [6.10]. EUREKA uses it for the pro-factory umbrella project [6.11] and the new cluster: ManuFuture industry.

Several hundreds of proposals, based on SI to maximise the research-market value chain, have been submitted to PAs calls.

Beside the results previously presented, the ManuFuture framework for implementation has brought about the development of best practices at E&RTD&I entities level.

ManuFuture SI recommendations for the mutation of the E&RTD&I system are being adopted and used by stakeholders within intelligent K-Triangles to generate and test new IK-T configurations. Some preliminary best practices are presented in the next paragraph.

They should contribute to improve the education-RTD-innovation-market value chain. Two ongoing pilot initiatives concerning education and complying with ManuFuture recommendations are reported below.

Learning Factory

The objective is to promote knowledge, competences and best practices for advanced industry, by integrating learning, research, innovation activities, through a context-aware virtual factory for collaborative learning. This is needed (see figure 6.6) in order to promote and support HAV knowledge-based, competitive sustainable manufacturing industry.

Learning Factory

E-PROCESS	• Increasing relevance of industrial engineering • Developing new and innovative models, techniques and methods for collaborative learning • Developing a collaborative learning enviroment by extend adapt and integrate the already state-of-art collaborative systems, architectures, tools and frameworks
PROFESSIONAL SKILLS	• Knowledge Modeller and Storer • The Knowledge Asseser
CHALLENGES	• Dynamic collaborative e-learnig enviroment in industrial engineering • Creating virtual classrooms or virtual centers for students • Digital tools and simulators able to be used anywhere and anytime • E-training enviroments context-aware
DRIVERS	• Collaborative design and engineering, simulation, virtual manufacturing, digital and virtual factory • Learning of advanced skills the acquisition of specialized knowledge, by using specific methods and tools • Work collaborative, remotely and sometimes while moving

Fig. 6.6. Learning Factory Pilot Initiative [6.12]

Learning Factory

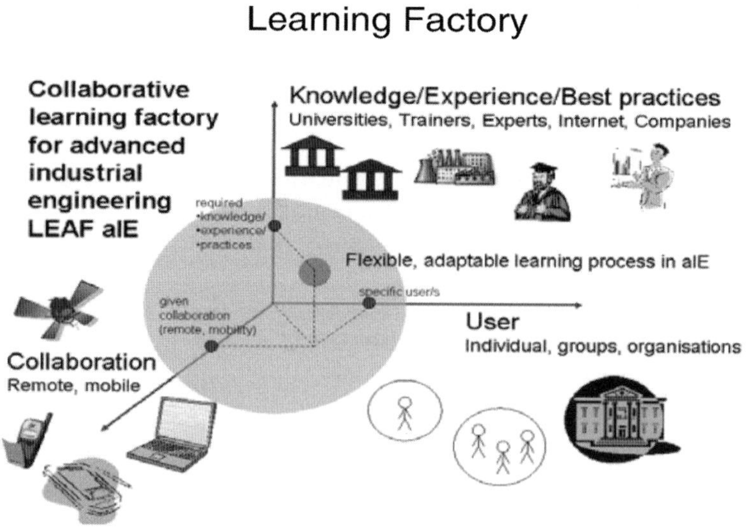

Fig. 6.7. Learning Factory Pilot Initiative [6.12]

Teaching Factory

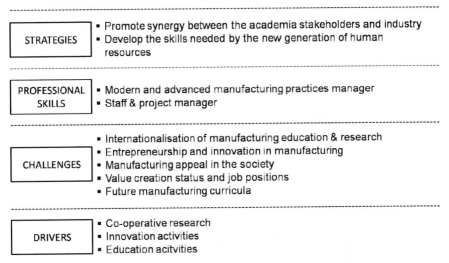

STRATEGIES	• Promote synergy between the academia stakeholders and industry • Develop the skills needed by the new generation of human resources
PROFESSIONAL SKILLS	• Modern and advanced manufacturing practices manager • Staff & project manager
CHALLENGES	• Internationalisation of manufacturing education & research • Entrepreneurship and innovation in manufacturing • Manufacturing appeal in the society • Value creation status and job positions • Future manufacturing curricula
DRIVERS	• Co-operative research • Innovation activities • Education acitvities

Fig. 6.8. Teaching Factory Pilot Initiative [6.13]

The *Teaching Factory*
a catalyst for academia – industry interactions

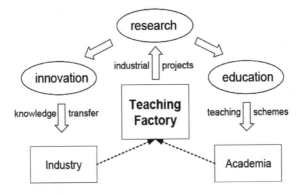

Fig. 6.9. Teaching Factory Pilot Initiative [6.13]

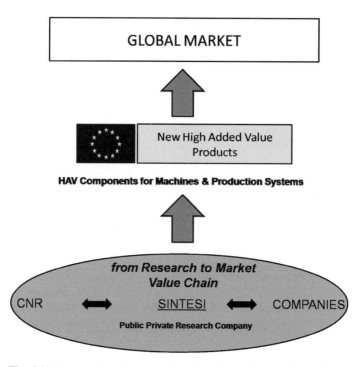

Fig. 6.10. National Joint Research-Industry Pilot Initiative: Sintesi [6.14]

The objective is to seamlessly integrate education, research, innovation activities within a single initiative, to develop competences and skills needed to promote and support future perspectives of a HAV, knowledge-based, competitive sustainable manufacturing industry.

Two ongoing pilot initiatives, concerning E&RTD&I and complying with ManuFuture recommendations, are reported below.

- A national joint research-industry pilot initiative: Sintesi (figure 6.10). The objective is to develop – for the European Market and further – K-based components to be embedded in machines and production systems and hence increase HAV of the latter. The National Research Council (CNR) of Italy and ten companies producing machinery and systems have set up a public-private research company (Sintesi SCpA). The HAV chain, going from scientific research to industrial innovation and market, has been implemented and is active within Sintesi premises, thanks to a hundred especially trained bright young industrial researchers: a ManuFuture goal. They have already developed components and machines far beyond the state of the art and taken them to the market [6.14].
- A European Joint Research-Industry Pilot Initiative: Synesis (figure 6.11). The objective is to contribute to the achievement of ManuFuture goal: the factory as a product, for the European and global market. CNR and FHG (Fraunhofer Gesell-schaft) have set up a public private research company (Synesis) in co-operation with industries.

This will carry out research-based innovation and related education and training within locations in Italy and Germany, thus implementing a European value chain going from scientific research to industrial innovation and market. Both institutions will contribute to the pilot initiative in terms of unique pilot plants, acting as laboratories and advanced enabling technologies. Synesis will support the emerging EMRI (European ManuFuture Research Institute) and is open for future European partnerships.

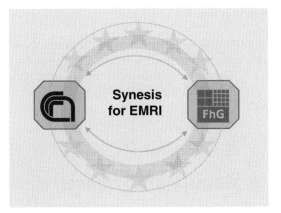

Fig. 6.11. A European Joint Research-Industry Pilot Initiative: Synesis [6.15]

6.2.2 ManuFuture Pilot Actions

Due to the size and complexity of the transformation process, that concerns industry and the E&RTD&I system, governance and effectiveness, the pursuit of CSM requires pilot and leading actions.

A Manufacturing JTI is currently being considered. Its objective is the implementation of manufacturing enabling technologies of the future, based on the ManuFuture SRA.

The JTI is a new instrument of the 7th FP for setting up long-term public-private partnerships, focusing on areas where research and technological development can contribute to European competitiveness and quality of life. By joining forces and pooling resources, industry can accomplish far more than by doing it alone.

Further "Lighthouse Projects" are being considered, operating in domains where Europe may strengthen its leadership and accelerate the transformation process. They concern from SI implementation, to developing and assessing new FW features, to inter-European and global co-operation, to monitoring the transformation in progress.

The first lighthouse project launched concerns European production systems. Its rolling master plan covers from medium to medium-short-term horizon and refers to FP7, Eranet, EUREKA: i.e. the ongoing Pro-Factory Umbrella Project and the starting ManuFuture Industry (MF.IND) cluster project [6.16]. It is worth noticing that a European platform, as ManuFuture, together with a Eureka cluster such as ManuFuture Industry, may be a coalescence centre for a JTI.

Acting within ManuFuture framework and using ManuFuture SI, MF.IND focuses on the dual role of HAV, K-based European production systems (see figure 4.15) :i.e. as products to be sold globally, by European companies, and as processes and enabling technology, that can contribute to the shift of the European manufacturing 'fabric' towards CSM.

Networks of high-tech technology providers and producers are being set up to carry out from E&RTD&I activities to industrial validation and further. Activities will concern from knowledge and standards, to enabling technologies, business models, and organisations, products and services for different market segments, processes for different sectors.

The major impact will be on the global competitiveness and sustainability of technology and machinery and systems suppliers, systems integrators, final products producers. Co-operation at European, international, global level will be pursued. As a cluster, MF.IND (figure 6.12) will transform available SI into the specific SI, necessary to run the E&RTD&I activities.

Fig. 6.12. The EUREKA Cluster ManuFuture Industry, MF.IND ©Sintesi [6.16]

This will be complemented by a monitoring process, concerning ongoing projects as well as ESET context evolution. Thus, the governance of MF.IND will be insured. Time constants should be minimised.

More than 40 industrial partners and ten research institutes and centres, from 15 countries, make MF.IND partnership.

This is a KIC that will be organised as an association. In addition to carrying out RTD&I activities the MF.IND cluster project and its KIC will develop and test new CSM-oriented E&RTD&I pilot initiatives and structural models: i.e, HAV IK-Ts. Governance methods and tools, preparing for EIT and JTI, and co-operation models, from European to global level, will be developed and tested. MF.IND is expected to launch projects for 400 MEURO.

References

[6.1] Jovane F.: Towards Knowledge and Innovation Communities. In: ITIA-CNR- ManuFuture Conference 2007, Porto (December 2007)
[6.2] Commission Staff Working Document – Accompanying the Green Paper - The European Research Area: New Perspectives (April 2007)
[6.3] Research Advisory Board Final Report – Energising Europe's Knowledge Triangle of Research, Education and Innovation through the Structural Funds (April 2007)

[6.4] ManuFuture Platform — Catalogue of ManuFuture National and Regional Technological Platforms (2008)

[6.5] European Commission – Improving knowledge transfer between research institutions and industry across Europe: embracing open innovation – Implementing the Lisbon agenda (April 2007)

[6.6] Jovane, F.: ManuFuture for the Manufacturing Sector evolution. In: ITIA-CNR-ManuFuture Italian National Conference Milan (April 22, 2008)

[6.7] Etzkowitz, H.: The Triple Helix of University - Industry - Government - Implications for Policy and Evaluation – Science Policy Institute - Working paper (2002)

[6.8] Westkämper, E.: Private Archive

[6.9] Jovane, F., Yoshikawa, H., Alting, L., Boer, C., Westkämper, E., Williams, D., Tseng, M., Seliger, G., Paci, A.M.: The Incoming Global Technological and Industrial Revolution - towards Competitive Sustainable Manufacturing – CIRP Key Note Paper (to be published)

[6.10] European Commission – Seventh Framework Programme Vision Papers (2007)

[6.11] EUREKA Umbrella 4090 PRO-FACTORY The Scope of Pro-Factory 2007 - Proposal by Hubert van Belle (June 2007)

[6.12] Constantinescu, C., Hummel, V., Westkämper, E.: Collaborative Learning Factory for advanced Industrial Engineering, LEAF aIE (2006)

[6.13] Chryssolouris, G., Mavrikios, D.: Education for Next Generation Manufacturing – Cirp (2005)

[6.14] Martana, R.: Sintesi. Research, Training and Innovation for the Machinery and System Sector C – Porto 2007 (2007)

[6.15] Jovane, F.: CNR Research and Innovation in Manufacturing Systems in Lombardia: from regional to inter-regional and European successful activities. In: Bi-regional Workshop Lombardia -Baden-Württemberg – Stuttgart (January 2007)

[6.16] EUREKA Cluster E!4456 ManuFuture Industry (MF.IND), White Book and Roadmaps (April 2008)

7

Final Conclusions

Knowledge-based Competitive Sustainable Manufacturing (CSM) must be developed and implemented in Europe within a framework of growing global co-operation. Pursuing it must be seen as the new *European Technological and Industrial Revolution* for global competitiveness and sustainability. It is the main enabler of sustainable development,

This will help European manufacturing industry retain and improve its role as wealth generator, job provider and resource user. More than 34 million people are employed in more than 2.230.000 manufacturing enterprises in 23 industrial sectors. Manufacturing turnover accounts for more than 6.300 BEURO, 55% of the European GDP with an added value of 1.630 BEURO. A further 60 million people are employed in manufacturing related services. Nearly 500.000 people are engaged in research, technological development and innovation, and related education, within university, research institutes and centres as well as industry.

Europe still leads the global trade market, but competition from advanced and emerging countries is undermining its position.

European industry must attain the leading position in competitiveness and sustainability. To this end European industry should undergo a transformation towards HAV, K-based. This must be enabled and supported by the E&RTD&I (the K-triangle) system that, in its turn, must become more robust and effective, competitive and sustainable, and increase its global reach. Both industry and the E&RTD&I system must undergo – simultaneously and integrated – transformation processes, within two interlinked domains, getting global: i.e. the internal and global market, and EMIRA, the European manufacturing innovation and research area.

As presented in this book, the ManuFuture reference model of basic activities and key pilot actions shows the stakeholders – from policymakers, to PAs, at EU, national and regional level, and financial institutions; to industry, university and research institutes and centres – the way to act within the EU and at a global level to pursue CSM. Further, the work done and in progress by the ManuFuture platform makes the requisite Strategic Intelligence (SI) available to the stakeholders and gives the ManuFuture framework to guide its implementation.

Within the ManuFuture framework fast and effective transformation may take place if implementation of the consequences of this SI is made by all stakeholders. Thus, PAs, at all levels, should launch E&RTD&I programmes and financial institutions should make funds available for investments. Industry, university and research institutes, acting within intelligent K-Triangles, should conceive and launch E&RTD&I projects and create HAV research-innovation-market value chains (figure 7.1).

Pilot actions are now being launched. The EUREKA cluster ManuFuture Industry has been initiated, as a part of a ManuFuture 'Lighthouse Project', addressing European production systems.

Pilot actions will have a leading role and significant impact. However as has been presented, the new *European Technological and Industrial Revolution* for global competitiveness and sustainability will require a massive and co-ordinated activity. The ManuFuture Porto Manifesto (see annex I) supplies the headlines for the necessary actions.

The ManuFuture Knowledge Innovation Community (KIC) of several thousand committed players acting within the ManuFuture framework, can turn these into reality and success!

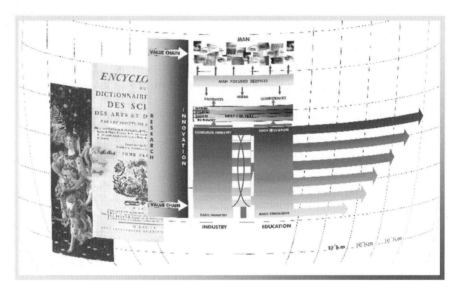

Fig. 7.1. ManuFuture 2003 Conference. Research-Innovation-Global Market-Value Chain

General References

Ayres, R.U. (ed.): International Institute for Applied Systems Analysis: Computer Integrated Manufacturing. IIASA Computer Integrated Manufacturing Series, vol. IV. Chapman and Hall, London (1992)

Ayres, R.U., Warnecke, H.-J. (eds.): International Institute for Applied Systems Analysis: Computer Integrated Manufacturing. IIASA Computer Integrated Manufacturing Series, vol. II. Chapman and Hall, London (1991)

Ayres, R.U.: Computer Integrated Manufacturing. IIASA Computer Integrated Manufacturing Series, vol. I. Chapman and Hall, London (1991)

Bullinger, H.-J., Warnecke, H.-J., Westkämper, E.(eds.), Niemann, J., Balve, P., Bauer, S., Gerlach, G. (reds.): Neue Organisationsformen im Unternehmen Ein Handbuch für das moderne Management. 2., neubearb. u. erw. Aufl. Springer, Berlin (2003)

Club of Rome, The Limit to Growth. Rome (1972)

Craighill, A.L., Powell, J.C.: Life cycle assessment and economic evaluation of recycling: a case study, Resources. Conservation and Recycling 17(2) (August 1996)

EC – DG Research; The Future of Manufacturing in Europe 2015-2020: The Challenge for Sustainable Development (FuTMan) (2002)

EC Conference ManuFuture 2003. European Manufacturing of the Future: role of research and education for European leadership., Milano (December 1-2, 2003)

Ford, H.: Today and Tomorrow, Reprint Edition. Productivity Press, Portland (1999)

Handbook of production management methods / Gideon Halevi. Butterworth-Heinemann, Oxford (2001)

Imai, M., Kaizen, G.: A Commonsense, Low cost Approach to Management. McGraw-Hill, New York (1997)

Intelligent Manufacturing Systems; Impact Report – History and Achievements of Phase 1 (2005), https://www.ims.org

Jeffrey, K.: Liker: The Toyota Way: 14 Management Principles from the World's Greatest Manufacturer. McGraw-Hill, New York (2004)

Jha, N.K. (ed.): Handbook of Flexible Manufacturing Systems. Academic Press, San Diego (1991)

Knot, J., Marjolijn, C., van den Ende, J.C.M., Vergragt, P.J.: Flexibility strategies for sustainable technology development. Technovation 21(6) (June 2001)

Ayres, R.U. (ed.): International Institute for Applied Systems Analysis: Computer Integrated Manufacturing. IIASA Computer Integrated Manufacturing Series, vol. IV. Chapman and Hall, London (1992)

Ayres, R.U., Warnecke, H.-J. (eds.): International Institute for Applied Systems Analysis: Computer Integrated Manufacturing. IIASA Computer Integrated Manufacturing Series, vol. II. Chapman and Hall, London (1991)

Ayres, R.U.: Computer Integrated Manufacturing. IIASA Computer Integrated Manufacturing Series, vol. I. Chapman and Hall, London (1991)

Bullinger, H.-J., Warnecke, H.-J., Westkämper, E.(eds.), Niemann, J., Balve, P., Bauer, S., Gerlach, G. (reds.): Neue Organisationsformen im Unternehmen Ein Handbuch für das moderne Management. 2., neubearb. u. erw. Aufl. Springer, Berlin (2003)

Club of Rome, The Limit to Growth. Rome (1972)

Craighill, A.L., Powell, J.C.: Life cycle assessment and economic evaluation of recycling: a case study, Resources. Conservation and Recycling 17(2) (August 1996)

EC – DG Research; The Future of Manufacturing in Europe 2015-2020: The Challenge for Sustainable Development (FuTMan) (2002)

EC Conference ManuFuture 2003. European Manufacturing of the Future: role of research and education for European leadership., Milano (December 1-2, 2003)

Ford, H.: Today and Tomorrow, Reprint Edition. Productivity Press, Portland (1999)

Handbook of production management methods / Gideon Halevi. Butterworth-Heinemann, Oxford (2001)

Imai, M., Kaizen, G.: A Commonsense, Low cost Approach to Management. McGraw-Hill, New York (1997)

Intelligent Manufacturing Systems; Impact Report – History and Achievements of Phase 1 (2005), https://www.ims.org

Jeffrey, K.: Liker: The Toyota Way: 14 Management Principles from the World's Greatest Manufacturer. McGraw-Hill, New York (2004)

Jha, N.K. (ed.): Handbook of Flexible Manufacturing Systems. Academic Press, San Diego (1991)

Knot, J., Marjolijn, C., van den Ende, J.C.M., Vergragt, P.J.: Flexibility strategies for sustainable technology development. Technovation 21(6) (June 2001)

Porter, M.: Competitive Advantage (1985)

Monden, Y.: The Toyota Production System: An integrated Approach to Just-In-Time, 3rd edn. Engineering and Management Press, Norcross (1998)

Sarkis, J.: Manufacturing strategy and environmental consciousness. Technovation 15(2), 79–97 (1995)

Spur, G. (ed.): Fabrikbetrieb. München, Wien: Hanser (Handbuch der Fertigungstechnik 6) (1994)

Ohno, T.: The Toyota Production System: Beyond Large-Scale Production. Productivity Press, Portland (1998)

Taylor, F.W.: The Principles of Scientific Management. Harper & Row, New York (1911)

The Future of Manufacturing in Europe. 2015 -2020: The Challenge for Sustainability DG research

United Nations – Report of the World Commission on Environment and Development General Assembly Resolution 42/187 (December 11, 1987)

Warnecke, H.J.: The Fractal Company – a Revolution in Corporate Culture. Springer, Berlin (1993)

Womack, J.P., Jones, D.T., Roos, D.: The Machine that Changed the World: The Story of Lean Production. Harper Perennial, New York (1991)

Womack, J.P., Jones, D.T.: Lean Thinking: Banish Waste and Create Wealth in your Corporation, Revised and updated, 2nd edn. Simon & Schuster, New York (2003)

Zhang, H.C., Kuo, T.C., Huitian, L., Huang, S.H.: Environmentally Conscious Design and Manufacturing: A State-of-the-Art Survey. Journal of Manufacturing Systems 16(5) (1997)

ANNEX I

Manufuture Porto Manifesto

Context

Competitive and Sustainable Development is emerging as the compelling global strategic vision to be deployed as to meet the economical, social, environmental and technological challenges that today's European societies are facing.

Manufacturing, generating wealth and jobs by fully exploiting know-ledge and resources, is the fundamental enabler and sustainer of Europe's Competitive and Sustainable Development. Manufacturing in Europe provides presently 41.5 % of the added-value (over €1,535 million) and 30.4 % of the employment (34 million people), with each job at the factory floor generating two other jobs in services. Manufacturing, as the heart beat of growth and development, must become increasingly High-Adding-Value, competitive and sustainable, by building on competences and knowledge coming from high education and R&D.

Transformation of industry, as well as transformation of the High Education and Research and Innovation Systems, may be considered as the forthcoming European Scientific, Technological and Industrial Revolution, that is necessary for Europe to lead at global level.

The ManuFuture platform has developed the strategic intelligence – encompassing Vision, Strategic Research Agenda and Industry Roadmaps – enabling the stakeholders to power and drive such a revolution. The stakeholders involved at the European level range from industry to academia and research institutes, from public authorities to financial institutions.

The common understanding that has been gained since the Platform was launched in 2003 until today has been validated by nearly 400 stakeholders in December 2007 and has been put together in the present ManuFuture Porto Manifesto.

This manifesto is also meant to be an answer of the stakeholders to the EC Recommendation "Integrated Guidelines for Growth and Jobs (2008-2010)", which has been proposed for a Council Decision on the guidelines for the employment policies of the

Member States (under article 128 of the EC Treaty) at the 14th December 2008 meeting of the Portuguese Presidency of the EU. The alignment with many of the microeconomic guidelines of the Recommendation – namely those regarding R&D investment, innovation deployment, industrial base competitiveness, sustainable growth and better regulation – is of paramount importance. Some of the critical issues targeted by the employment guidelines – dealing with improving quality and productivity at work, investment in human capital and adapt education and training system in response to new competence requirements – are also present in the Manifesto.

The aim of the ManuFuture Porto Manifesto is nevertheless to go one step further and driving strategy into action. This is why it is structured into very concise action leads for each challenge selected as priority.

Action Leads

1. **Leading sectors** of the European industry need to further reinforce their competitiveness through increased investment in R&D
   Lines of action:
   1.1 – increase awareness on the economic impact of R&D investments and their leveraging effect on future sustainable competitive advantage
   1.2 – tackle large investments in R&D with new funding models, building on R&D tax incentives (a proven risk sharing mechanism) as well as on novel financial tools, such as the *Risk Sharing Financing Facility* recently launched jointly by the EC and the European Investment Bank

2. **Mature manufacturing sectors** are presently in crucial and urgent need to add value and decrease costs by embedding design and technology, as to compensate for the fierce competition from the emerging economies.
   Lines of action:
   2.1 – stabilise mature sectors through medium-long term strategies supporting the development of R&D-based competencies in high-value complex manufacturing technologies which are unique and difficult to replicate outside of Europe
   2.2 – improve SME access to the best R&D resources and institutions

3. It is crucial for Europe to intensify the exploitation of leading-edge science, technology and knowledge, creating wealth and highly-qualified jobs, specially in **emerging sectors** which will foster future high-value markets
   Lines of action:
   3.1 – invest in novel R&D intensive businesses built on disruptive technologies, such as nanotechnologies
   3.2 – improve innovation management while shifting from cost to high value adding
   3.3 – promote a new culture of risk acceptance and dealing with failure as to foster entrepreneurship

3.4 – strongly increase European <u>expertise in venture capital investment</u> (due-diligence of ideas, technologies, market evaluation, business and management skills appraisal)

3.5 – leverage the <u>globalisation of early-stage high-tech companies</u>

4. With <u>highly qualified **human resources**</u> at all levels being the <u>differentiating asset,</u> European manufacturing has to regard expert labor as the most critical sustainable resource and has to contribute to offset USA advantage (Europe has only 5,5 researchers per 1.000 inhabitants against 9 in USA) and to revert the present brain drain to USA

Lines of action:

4.1 – define a <u>manufacturing industry education and training agenda,</u> articulating with universities and poly-technical schools on the requirements for advanced research-based specialised training capable of providing crucial domain-specific skills and fostering new teaching principles and new engineering disciplines in under-graduate courses, post-graduate industrial training, industry theses, cooperative R&D initiatives, life-long training, etc.

4.2 – increase <u>employment of researchers in industry</u>, while anchoring company strategic research and innovation activities on nuclei of internal staff able to network and access the external competencies needed

4.3 – participate in the <u>university modernisation process</u> that is on-going in many European countries towards greater autonomy and increased responsibility in responding to societal needs, namely through the active involvement in public-private partnerships for the governance of research and higher education institutions

5. The <u>need to keep manufacturing operations in Europe</u> calls for **industry transformation** as well as for the **improvement of existing industrial units**, as to ensure strong cost reductions, increased flexibility and smaller response times while keeping high standards in product quality with increasing complex novel products

Lines of action:

5.1 – strong investment is needed in <u>green-field projects</u> and/or in <u>revamping</u> existing factories

5.2 – generalise the use of <u>benchmarking of industrial best practices</u>, specially in sectors that are under strong pressure from low wage countries, leading to both technological an organisational enabling changes

5.3 – <u>manage increased product complexity</u>, by spreading subcontracting/outsourcing and building collaborative networks over the complete supply and value chain

5.4 – enhance process flexibility through the use of state-of-the-art technology

5.5 – strongly increase cooperation with R&D&I entities

5.6 – make governments constantly aware of the extreme burden of the high costs of the good infrastructure in western countries which results in an excessive (industry) taxation

6. Visionary concepts like *Factories as a Product* – to be sold globally by European companies – or *Knowledge-based Factories Made in Europe,* should <u>aggregate</u>

and align the vast and rich European know-how in concrete initiatives paving the way for the transformation of the industrial fabric towards the **European leadership in sustainable manufacturing**
Lines of action:
6.1 – enable digital production
6.2 – control and install disruptive technological processes
6.3 – design and deploy a competitively sustainable European Production System

7. The quest for **regulation**, a differentiating attitude of European governments and institutions, has to be used to the advantage of European manufacturing
Lines of action:
7.1 – to facilitate and stimulate new business development – as in the case of energy, environment, transport or health markets – towards future areas of European leadership
7.2 – to encourage and not hamper entrepreneurship and to safeguard productivity from useless bureaucracy

8. A significant amount of European **R&D, engineering and design output** is often left **unprotected** to copy by Asian competition
Lines of action:
8.1 – reinforce awareness on IPR generation and on other IPR related issues (such as protection, licensing and commercialisation)
8.2 – improve existing patenting practices in Europe
8.3 – create a simpler and more effective European framework for IPR management

9. Promote the **image of manufacturing** near the population as to attract the interest of young educated and skilled people
Lines of action:
9.1 – promote public awareness of the relevance of manufacturing science and technology in the products, systems and services which build the backbone of today's society
9.2 – demonstrate the need for science-based innovation in competitive (creating jobs and wealth) and sustainable (environmentally friendly) manufacturing
9.3 – advertise creativity, sophistication and added-value in jobs in modern European manufacturing

10. Differences across **european regions** should lead to **specific positioning and strategies** by each region as to maximise their capabilities and resources
Lines of action:
10.1 – develop and/or reinforce regional innovation platforms
10.2 – promote most developed regions as benchmarks and engines to other regions and build networks across regions with specific objectives of mutual benefit, such as competitive cost supply chains to large OEM's
10.3 – build a European competence map in manufacturing science and technology organising the tremendous potential of universities and R&D institutes across Europe for the use of industry and other stakeholders. The universities and

research centres are expected to describe and advertise their competencies in research and education as if they were tier-1 suppliers in a supply chain. A *European Manufacturing Engineering Curriculum* (following the example of the GME, Global Education in Manufacturing initiative) should be agreed by leading European universities.

10.4 – increase emphasis on knowledge sharing and research collaboration across regions, institutions and industries, developing new mechanisms of knowledge flow

11. The seamless **integration of EU, national and regional policies** – crucial for stimulating targeted, problem-solving R&D and innovation efforts – calls for the urgent attention of the EC, national governments, regional bodies, industry associations and academia

Lines of action:

11.1 – initiate an appraisal of research strategies in political bodies to encourage "human driven innovation", instead of an overemphasis on mere technological innovation

11.2 – align efforts to overcome fragmentation in EU R&D efforts, taking full advantage of European Technology Platforms, Joint Technology Initiatives, etc., paving the way to an "European Manufacturing Technology Institute"

11.3 – simplify procedures and ease the burden of bureaucratic requirements to be met by projects, initiatives and stakeholders in EU Framework Programs, ERDF projects, as well as in national and regional support measures

Porto, on the 4th December 2007

The ManuFuture *Platform Stakeholders*

Annex II: Proactive Initiatives Defined by the Leadership Consortium

Contents

Proactive Actions Priorities

Take into account the strategic road from competition to global leadership and many experts opinion

L   = Large, strategic importance
M   = Medium
S   = Small

Time Scale

ST   = Short-term   1-3 years
MT   = Medium-term   3-5 years
LT   = Long-term   5-10 years

1 New Business Models

1.1 Beyond Lean Manufacturing

M
ST

The set of methodologies of lean management defines the management system in series production. The transformation and adaptation of these methodologies towards higher complexity, higher variants and changeable manufacturing aim at activating productivity potential in series and small batch production.

Lean management methodologies, which have been successfully implemented in automotive and supplier industries, are quasi standards of today's efficient manufacturing. The competitiveness under the European conditions of high wages and changing markets and technologies revealed that lean management is a prerequisite but cannot reach the cost advantages in general. Lean management methodologies are required for customised manufacturing.

1.1.1 Preventive Quality Management

S
ST

The quality and reliability of products, services and industrial operations (business processes) are preconditions for High-Adding-Values and the growth of demanding manufacturing sectors. Following the trends towards customised products and build to order strategies in manufacturing, new and efficient methods are required in all manufacturing sectors, in order to assure quality and reliability in early phases of products. Preventive quality management approaches are required. They include innovative methodologies for the introduction and management of the life cycle of new generations of products, such as those including mechatronics and intelligent products.

There should be a focus on the design of quality and reliability and the design process should include the ability to forecast the utilisation and life time of complex products. Methodologies have to take into account the development of products by co-operative and networked engineering, the capability of the design, manufacturing and measurement processes and the influences of manufacturing technology choices.

Proactive Initiatives   **Priority**

Several main results can be envisioned: (1) Basic methodologies for preventive and life cycle-oriented quality management including improved design (2) Increased reliability of complex products (3) Development of an European product documentation system (4) Reduced losses caused by quality problems (5) trans-sectorial implementation for pushing High-Adding-Value in the manufacturing of complex products.

1.1.2 Manufacturing Fitness, Balancing Reactivity and Efficiency

M

Industrial manufacturing is oriented towards achieving the main objec-   ST
tives of time, cost and quality. Industrial paradigms strictly follow conventional paradigms like balancing the capacity or management of resources with high rates of utilisation. The next-generation manufacturing is characterised by customisation which reduces the lot sizes and increases the variants and specific products, manufactured in a short time. Another specificity of next-generation manufacturing is the increasing complexity of products. This challenge can be overcome through actions like the development of methodologies for future manufacturing management, based on new and innovative paradigms, mainly aiming at achieving following goals: single customer order, flexible work time and high reactivity, especially in the sectors of customised products. Market influences and changing customised products by reducing the lot sizes to 1 with increasing inefficiencies, requires new methodologies for the balancing of capacities in turbulent markets.

These are intended to support the:

- Flexibility of resources and flexible work management
- Situation, based on balancing of the capacity load
- Self-organisation and self-optimisation, self-controlling with autonomous work groups and business units
- In-situ Management
- Learning organization

1.2 Survival Strategies in a Turbulent Industrial Environment and High-Adding-Value in the Life Cycle

L

ST

The development of the market depends mainly on economic factors. In many industrial sectors and especially in investment sectors, the normal good cyclic fluctuations with strong ups and downs represent main factors of market turbulences. Many enterprises are not capable to remain competitive or at worst, they do not survive in these cyclic

Proactive Initiatives   **Priority**

phases. Other challenges are issued by market strategies of competitors, which operate in areas characterised by lower wages and higher flexibility of the work force (hire and fire). European human-oriented culture and social standards on the one side and high fix costs caused by capital intensive production on the other side highly reduce the chances of surviving. European manufacturing industries, operating under the increasing pressure of making short time profits, represent another turbulence factor. All these factors reduce the employment and sustainability of many industrial sectors. The current economic models are mainly oriented towards growth and are following the preventive strategies coming from product innovations. There is a requirement for manufacturing strategies and methods for:

- Balancing the load in mid-term cyclic markets
- Overcome critical short-term situations
- Dynamic forecast
- Adaptation of fix costs (dynamic systems)
- Dynamic work force models
- Financing of critical phases.

These strategies represent the main directions on which research and technological development activities have to be oriented, with high priority on short term horizons.

1.2.1 Transformation Management Strategies for Survival and Success in Turbulent Environments

M
ST

Manufacturing enterprises are influenced by multiple dynamic external factors concerning the products behaviour in global markets, the strategies of competitors, the regional level of wage and reward systems including management of employees healthcare cost, regional infrastructure, the pace of technical innovations, the financial requirements of the investors and the financial constraints of operations, the robust supply of materials and components. Internal business factors such as qualification and capability of employees and the management, the demands and systems required by different customers, the utilisation of resources and the capability of processes as well are influencing factors. The enterprise environment is tough and turbulent. Only those enterprises can survive and be successful in this turbulent environment which are robust enough and have the capability to continuous adaptations and transformations.

These challenges are particularly acute for SMEs which operate in traditional and new technology sectors and do not have the scale and resources to address all the changes in their environment. The technical content of transformation management strategies for survival and

Proactive Initiatives	Priority

success in turbulent environments is the development of such strategies that recognise the evolution of the manufacturing business environment. Research challenges envisioned here include: identification of methods for small businesses aiming at recognising and responding to external threats; determining and assessing candidate business models for SMEs to assist their survival and transformation; defining mechanisms that allow SMEs to take similar advantages from manufacturing in the enlarged Europe in comparison with those accessible to large companies; determining the required competences and mechanisms for SMEs to form co-operative transient business networks to increase their scale and to respond to opportunities; determining and disseminating best practice survival and transformation mechanisms developed by successful SME businesses or agencies; generating tools and techniques that support transformation; and the development of economic regulations and financial instruments that support transformation.

1.2.2   New Product and Process Life Cycle-Oriented Strategies

M
MT

The new paradigm of manufacturing is oriented towards the optimisation and value creation of products along their whole life. This assumes the prediction and understanding of the future user requirements and design of products (customisation), the manufacturing, product-near services and end-of-life. The global market is registering an increasing demand of customised products, which have a short delivery time: in parallel, a continuous shift of business is taking place towards the development of a new and innovative system of products and services, capable of fulfilling specific user demands. Under this perspective, the analysis and the best orientation of the product life cycle, to achieve maximisation of potential and of related business opportunities, are crucial elements that have to be exploited. Furthermore, the integration of new technological developments in products and consequent modifications in the production process are driving the manufacturing sector towards complex and articulated dynamics that require strategic intelligence and a new role of workers. To properly face these challenges, the production process shall be analysed and optimised in its structural and functional aspects, considering the total life cycle, in order to identify qualitative aspects, which have to be enhanced, new performance factors, networked integration and interaction aspects, and environmental factors which have to be achieved.

Innovative and new combinations of the following elements – customisation – production – service – end-of-life, including within each element the solution of "what, how and when" in order to introduce knowledge, know-how and technologies, represent valuable solutions to the above stated challenges.

Proactive Initiatives   **Priority**

Several main development issues and targets at network and factory level have to be mentioned:

- Methodologies for regulating the new paradigm with strategic intelligence on markets, products, technologies and human resources (success stories, best practices, innovation laboratories)
- Methodologies and tools for understanding product/service potentials and for identifying new business opportunities with respect to new market/consumer demands, in terms of product enhancement and new consumer-oriented services
- Process architectural analysis, process functional analysis and performance calculation, multi level simulation, process performance optimisation product/service related
- Tool and methodological support for the integration of state-of-the-art industrial paradigms for manufacturing fitness, balancing reactivity and efficiency
- Tool and methodological support for integrating systems and processes of suppliers and customers capable of supporting continuous adaptation.

1.2.3 New Consumer-Oriented Business Models for Product Life Cycle

M
MT

Integrating systems and processes of suppliers and customers which are capable of supporting continuous adaptation to market needs are needed to strengthen the competitiveness of European manufacturing and logistics companies, facing new opportunities and threats due to continuing globalisation. The main development issues and targets are a further transition from products to solutions (services), improved and increased involvement of the consumer in more parts of the value chain and managing the consequences of the reduction of the vertical integration traditional in larger businesses. Further, environmental drivers and an increasing recognition that manufacturing businesses do not benefit sufficiently from the value that they create, emphasises the need for the development of a whole new life cycle-based business model that minimises environmental impacts while maintaining economic sustainability. The focus should be to resolve the challenges that need to be addressed to encourage businesses to work in open collaborations within a production and logistic network across the whole life cycle; the development of new cross-company business models, addressing aspects of cost, benefit and risk, sharing within coherent financial and intellectual property frameworks which give mutual benefits; and to define and create the supporting technologies and tools, which are necessary. These new business models should cope with new product concepts, also including aspects

Proactive Initiatives   **Priority**

of dismantling and recycling, and providing a set of services and functionalities, including a potential to upgrade products by "after sales".
The projects are expected to have a successful technology demonstration, technology transfers and training activities.

The performed research aims at achieving the following main results: an increase in added-value, productivity and economic sustainability in Europe through an industrial stakeholder involve-ment. New
industrial strategies will increase and sustain production capability and
capacity and responsiveness, improve manufacturability, quality
and reliability as well as decrease the consumption of raw materials
and energy.

1.3   Management of Complexity

L
ST

The manufacturing enterprises have to handle new demands arising
from the market, from new technologies or restriction by pollution,
from the society as a whole, becoming ever more complex in all aspects in the past years. The global economy and the socio-political
scenarios are increasingly more intricate, more fragile and difficult to
understand and manage.

Management of complexity is a very important issue for harmonising all these demands in a practical way, in a management strategy. It
implies identifying how complexity starts and works in the complete
manufacturing and distribution process.

The process of handling and managing complexity has to consider
the development of tools for:

- Complex visualisations
- Interdependent visualisations
- Scenario management, if –what analysis
- Complex simulations
- Reducing and optimising processes, interdependencies and
  connections.

The management of complexity will lead to successful manufacturing
operations, allowing manufacturing enterprises to be successful in
their markets and be more accurate and rational in their manufacturing
processes and their product range.

1.3.1   Innovation and Transformation Processes

S
MT

The transformation from basic research to application is essential for
the effectiveness of the research system and market success. Many
companies invest only a small part of their turnover in the development of new products, based on new results of basic research. It is

Proactive Initiatives   **Priority**

known that aggressive technology leaders combining product, production and marketing strategies are more successful than followers.
This is due to the structure of the research system but also a question
of reliability of the transformation process. The main constraints from
industrial perspectives are:

- Missing models for technology management and the integration of manufacturing strategies in the strategic planning
  processes.
- Missing methodologies for the integration of new technologies in the resource planning processes.
- Uncertainness about the potential and the effects in manufacturing.
- Missing experiences of practical points for reliability
- Process chains from research to practice

To overcome the above stated challenges and to accelerate the knowledge transfer for innovation purposes, research and technological
implementations aim at the development and evaluation of methods
for technology evaluation, the integration of manufacturing strategies
in business planning and on time information about knowledge of new
technologies. This includes the economic potential analysis of new
technologies by practical procedures: feasibility and reliability studies,
forecast and simulation, organisational integration of operations like
simultaneous strategy planning.

1.3.2   Change and Modification Management

M
MT

Due to increasing dynamics in the markets and decreasing product life
cycles production ramp-ups have to be performed both more often and
within shorter intervals in many high-volume industries. Especially
suppliers are facing time and cost problems in serving different original equipment manufacturers. In order to reduce development time,
avoid late product changes, and improve the co-ordination of development, engineering and production, the simultaneous engineer-ing
approach has been developed and diversely applied. As a result of
increasing product variety and augmenting fragmentation of the value
chain amongst many companies, the traditional SE-approaches are not
enough so that new approaches are required.

The results of a performed research in this field aim at expanding
original SE approaches in three dimensions. Firstly, a company must
be harmonised along its supply chain, supporting the fast and flexible
dissemination of technical changes in products over all participants of
the value chain. Secondly, while an increasing variety of products use
the same resources, solutions for handling changes shall be developed.

Proactive Initiatives   **Priority**

Thirdly, a company has to be able to propagate ramp-up activities throughout its global production network (e.g. start-up and launch of a pilot line at site A, afterwards shifting of series production to site B. Projects should address improvements especially for suppliers on all levels of the value chain, e.g. in automotive industry.	
## 1.4 Machine Tools and Centric Business Models Changes connected to the adoption of new business models for machine tools have special impact on the relationship between MT producers and MT users. In particular, the changes have impact on the responsibility of the results of operations, actually shared between builder and user. A builder is responsible for aspects which are mainly connected to the machine's safety and performance (precision, speed, reliability), while a user is mainly responsible for the parts which have to be realised (supplying man power, raw parts, tools, testing facilities, etc.). Adoption of new business models would affect this traditional model, in three aspects: 1. better definition of the 'grey areas' in machine tools of the producer/user relationship: some activities are not actually well covered by both of the two, such as machine installation in the job shop, its final testing, the definition of reliability-connected aspects (MTBF, MTTR, TCO); some performance measurements, etc. 2. NBM-oriented technical aspects: the adoption of a new type of relations between customer and MT producers, such as 'pay per use', 'Pay per part', etc. will move some responsibility of the production from MT users to MT producers, with consequences from a technical view point. 3. The more extreme possibility connected to the NBM and MT user/customer relationship is to transfer the complete responsibility of production to machine tool builders. Then, they will be responsible for parts of the design, technological cycles, machine tool definitions and set ups, raw parts procurements, man power; etc. The only difference between this model and the usual activities of Tier 1 is connected to the localisation of the production facilities in the end user plant. Then, the machine is owned by the producers, run with the producers' personnel, is maintained by the producer, discarded parts are just cost for MT producers; the end user just pays for finished parts. Taking into account the above mentioned aspects, a deep reconsideration of the roles and activities of MT builders and users, as well as of the technical, managerial, economical, financial, normative and legal aspects, might be taken into account.	M MT

Proactive Initiatives   **Priority**

The above mentioned aspects require a deep investigation and modifi-
cation of the traditional way of working, involving a wide number of
scientific and technical disciplines, as well as various industrial sectors. This will be done with the target to use the European machine
tool knowledge of manufacturers, acquired in decades of activity. This
knowledge can then be used to help the progressive loss of technical
specialisation coming from many industrial sectors, in which the focus
has moved from production to business, from commercial to financial
aspects, and to help to reduce the effects of progressive aging of population and of the reduction of interest for technical professions coming
from youth.

- Machine tools: the actual machine tool concepts might be re-
  considered to be adapted to different applicative contexts. In
  particular, they might be focused towards a higher flexibility
  and reconfigurability, in order to be adapted to various operational situations.
- Control and diagnosis: a wide use of ICT technologies can be
  useful to keep track of all the conditions of running machines
  located at customers' sites, as well as to define the number of
  manufactured parts and defect ratios, in order to allow the
  payment by the customer made under 'pay per …' strategies
- Project capabilities: Machine tools manufacturers might
  enlarge their knowledge base in order to cover the development of process and pro-duct designs. This would also lead to
  the integration of activities for a group of SMEs, in order to
  create a critical mass to face this challenge, helping to increase regional, national and European level collaborations
- Rules and standards: the actual standards and rules, connected to the use and testing of machine tools are focused on
  a traditional way of using machine tools. New concepts for
  testing might be developed in order to meet the requirements
  and avoid contrasts.
- Law aspects: the European and national legislations might be
  reviewed to cover the new kind of relationship that will derive from the new way of cooperation between MT users and
  producers, avoiding contrasts and respecting rights and com-
  mitments for all of them
- Financial and risk reduction aspects: the machine tools producer might deeply change its financial and risk management
  approach. In fact, in the traditional model, risk is evaluated ex
  ante, during the definition of contracts and with market
  analysis of potential customers, while financial aspects, connected to machine development and production, are covered
  with own capital or with bank financing or leasing, guaran-
  teed by the contract signed by the MT user. With new types

Proactive Initiatives   **Priority**

- of business relations, a deep involvement of new participants (such as insurance) will be necessary; again, the formation of consortia and association of companies will be necessary to create the critical mass to manage such aspects
- Intellectual property: new kinds of instruments for knowledge defence might be developed, in order to protect the rights of all actors involved in the NBM-based machine tools, giving also instruments that preserve information ownership during the design and commercial discussion phases
- Training activities and social aspects: the machine tool sectors will increase their attractiveness for young workers by multidisciplinary and will represent a very interesting playground for innovative forms of training, based on multimedia, virtual reality and other ICT-technologies.

The development of NBM-oriented machine tools is on the one hand, a multidisciplinary and transversal activity and, on the other hand, it will innovate one of the column of European industrial tissues, making it more fundamental for the worldwide manufacturing and ensuring a base of know-how, coming from the cooperation, based on co-design, co-technology and co-manufacturing with a wide number of customers, belonging to all manufacturing sectors worldwide.

# 2 Adaptive Manufacturing	
## 2.1 Assembly Systems	S/M ST

The assembly of customised and build-to-order products is one of the core competences of manufacturing. Short delivery times and an increasing complexity of products require high flexibility and permanent adaptation of the assembly systems. Hybrid systems with mixed automation, manual operations and assistance by robots are objectives of the technical development. Adaptation without losses of efficiency by set-ups can be realised by modularisation und plug-in technologies. The assembly execution system recognises the actual situation of the system, available resources and orders which are connected with links to PPC and MRP in real-time. The implementation of principles like self-organisation, self-learning and self-optimisation, which are based on the integration of multi-sensor/actor systems, leads to intelligent systems. But the variety of assemblies in customised manufacturing makes it necessary to change the operations in-situ between automation and human work.

2.1.1 Adaptive Technologies for Joining Processes

A specific element of this action is the integration of non-joining processes and assembly. The system has to be linked to the documentation of assembled parts and components as well as to measurements and physical tests. In-process measurement for quality reduces time and costs. Intelligent cognitive elements of adaptive assembly systems are the ability to learn, diagnostic features and in-situ simulations. It can be added by Internet information systems and human interfaces with voice processing and tactile feed-back.

By the employment of the above presented enabling technologies, several relevant results can be obtained: (1) Configurable systems for assemblies reduce costs and time even in customised manufacturing; (2) Front-ranking of the European manufacturers of assembly systems; (3) Adding-value in sectors of assembly suppliers, IT for manufacturing, control systems and services; (4) Benefits for the

The priority marking for section 2.1.1 is: M ST

Proactive Initiatives   **Priority**

users, mainly in sectors of automotive, electric and white products; (5) Leading the world market by the application of assembly technologies. ## 2.2 Flexible Machines for Rapid Reconfigurations The mechatronic components are widely used in end-products, for example in the automotive and aerospace industries. With increased autonomy they will offer a very effective way to configure robots and handling units. With increased precision and reliability (including fail-safe hard and software interfaces) they will become promising objects for the construction of rapidly reconfigurable manufacturing equipment, suitable to be used in a flexible, agent-based production environment. The main objective is to create radically new, self-adaptive machine structures with online self-optimisation, based on mechatronic modules. The knowledge-based and/or self-learning intelligent systems can feature multi-layer control, sensing and actuator structures with a high level of redundancy guaranteeing a high level of reliability and allowing optimal performance of a production system under different conditions. Innovation lies in moving from current 'assembled' sensor, actuator, and control system architectures to truly integrated mechatronic knowledge-based systems. Main development issues expected in this area are: 1. Development of tools for integrated optimised system configurations based on a mechatronic simulation with respect of the resulting performance (including damping characteristics, working envelope, etc.), 2. Development of adaptronic modules and their integration into intelligent manufacturing equipment: • active intelligent components (integrating sensors, actuators, control, mechanical structures), adaptronic modules and interfaces, MEMS, MOEMS) • enabling the production of micro systems, micro technologies (e.g. human machine interfaces dedicated to micro systems manufacturing, miniaturised manufacturing equipment…) • enabling advanced automated process control 3. Enabling knowledge-based, self-learning systems through the development of multi-layer controls and model based real-time compensation routines, embedding machining process knowledge 4. Development of flexible signal processing methods, and wireless communication mechanisms and flexible system busses with integrated power supply,	 M MT

Proactive Initiatives	Priority

5. Standardisation of mechanical, electrical and software interfaces.
6. Using the above, break the limits of conventional/existing manufacturing processes (machining, tooling, technologies), realising breakthrough of manufacturing methods and processes

Expected results are (i) tools and methods for mechatronic manufacturing systems and components modelling, set-up and use; (ii) demonstrating applications for mechatronic modules and their usage in machines and production systems.

2.2.1 Advanced Monitoring of Complex Manufacturing Systems

Priority: S, ST

Today, the monitoring of complex systems requires complex measurement and analysis functions. Algorithms use different sorts of knowledge but often depend on a regular up-date of a database or on the contact to a central knowledge base. Learning capabilities and the use of environmental data are limited.

Analysis systems for an advanced monitoring of manufacturing systems or complex products should work decentralised to evaluate the state of the monitored systems. Knowledge is essential for such tasks, so new concepts for knowledge acquisition and use are required. Innovative systems also use decentralised and distributed knowledge, new mechanisms for integration of heterogeneous data sources or completely new ideas concerning time and location of the knowledge-generation and use.

They utilise a wide variety of locally available data to give advice to the operator, including the remaining operating time, system degradation or time to the next service or repair. Also new concepts for user interaction to communicate the condition of the monitored system are applied to make operations more intuitive.

User interfaces should reduce complexity for the operator but simultaneously maintain the full extent of the system's control.

To cover new requirements or changed system environments, the functionality of the used measurement components can be changed by configuration and software adaptation; configuration of whole subsystems can be adjusted on the basis of experience and history.

2.2.2 Cost-Efficient Condition Monitoring Systems

Priority: S, ST

The research area should focus on developing systematic condition monitoring methodologies that are robust and cost-efficient, following possible directions:

Proactive Initiatives   **Priority**

- Introduction of physical models of the machine's behaviour in condition monitoring systems. This should reduce the required training effort, involved in state-of-the-art condition monitoring systems. The models should match the machine's operations at any state of degradation. To limit the modelling effort, the physical machine model is ideally composed of physical component models, delivered by the component supplier.
- Introduction of new sensors in production machines for condition monitoring should be minimised to reach cost-effectiveness. This can be achieved by advanced signal processing using existing sensors better.

The above mentioned scientific research goals can be achieved by:

- Combining information of multiple available sensors and controller signals (i.e. sensor fusion) new information can be obtained. A virtual sensor is realised like that.
- Also, sensors can be used more extensively, e.g. in transient modes of the machine. Hence more information can be obtained from already existing sensors.

2.2.3 Planning Tools for Open Reconfigurable and Adaptive Manufacturing Systems

S
MT

Process planning and process engineering are parts of the chain from design to manufacturing. Taking into account new solutions for configurable manufacturing systems, it is necessary to develop new and knowledge-based tools for the support of planning. The implementation of a knowledge system in this process can be realised by a platform for process planning which is integrated in the information and execution system of factories. Elements of this platform should be: actual data of the factories' resources and capabilities, modules and standards of processes, interactive and participative systems for process planning, design of specific equipments for time and cost calculation, programming of machines, robots and automated systems, communication and distributed work. At the horizon, virtual-real workplaces are able to optimise and monitor manufacturing, wherever in the world the processes are running. Acceleration of planning processes for fast and reliable manufacturing in all sectors of manufacturing should be achieved through implementing this technology.

2.3 Cooperative Machines and Control Systems

L
MT

The transformation of traditional production line concepts to non-hierarchical agglomerates of autonomous manufacturing units is a key

Proactive Initiatives	**Priority**
technology for the new European production. Research and development has to focus on the application of agent control technologies e.g holonic manufacturing systems, service-oriented control architectures for autonomous manufacturing components in the main European manufacturing domains. Novel approaches in these domains shall encompass the life cycle of the production systems from the development of generic manufacturing ontologies, methods and tools for the design of co-operative production systems, integrated engineering systems, monitoring and control systems, HMI for integration of human workforce, reconfigurability and behaviour. R&D projects should lead to generic system solutions and demonstrate applicability and current limitations in specific manufacturing domains. The research efforts will demonstrate the feasibility and technological advantage of the new European production in the core industrial domains. It is expected, that the results will stimulate important industrial innovations in production technology and enhance industrial work environments. The developed technology will drastically improve the international market position of European manufacturers in respect to reactivity on new manufacturing processes and product innovations.	
## 2.4 Intelligence-Based Process Capability Enhancement	M MT
Manufacturing processes are instable because of the high number of dynamic influencing factors (deviations of material, wear, dynamic of machines, etc). Manufacturing instabilities combined with the inaccuracy of measurement are compensated by the tolerance system. Tolerances are more and more reduced to guarantee the functions of products and to ensure the quality. Additionally, tolerances are defined for designing the end of manufacturing, but not the steps of processes (tolerance channel). To optimise the capability (cp) of processes today's post process measurement should be displaced by in-process or pre-process measurements. In order to control the process it is essential to integrate process models in the control system. This can include methodologies for signal analytics and machines – self-learning by implementation of cognitive systems.	

Specific features are the integration of sensors for measurable parameters under the specific conditions and systems for process control and monitoring. It is the objective to stabilise the process capability towards cp >2.0 even over a long time of usage of machines, taking into account the deviation and wear. All conventional and innovative technologies (casting, forming, cutting, joining, surface protection, laser-assisted technologies) are fields of this research towards intelligent manufacturing.

Proactive Initiatives   **Priority**

Main outcomes of the above presented enabling technologies represent the push of the manufacturing quality towards zero defects in
processes and process chains and realise intelligent self-optimising
manufacturing systems.

2.5 Manufacturing Control Systems for Adaptive, Scalable and Responsive Factories

L
MT

Modern manufacturing control systems must respond quickly to
continuous changes in the next generation responsive factory. With
traditional manufacturing control system programming, it is time-
consuming to make changes as a result of separate databases for the
programmable logic controller (PLC), Human/Machine Interface
(HMI), and supervisory control and data acquisition applications or
modules. New engineering approaches that ensure efficient, robust,
predictable, safe and secure behaviour for multi-scale, distributed,
scalable and responsive manufacturing and factories are required, as
well. The reconfigurability of software for current machine control
systems is very limited, although the concept of component-based
software integration has already been adopted in controller software
development. Specifically, the following limitations and then the challenges in current control software development practices hinder the
reconfigurability of manufacturing: 1) Application software is parti-
tioned and implemented with proprietary information, 2) Control behaviours of the software are either built inside the implementation and
hence, not customisable, or not modularised and associated with the
corresponding software components, 3) Software implementation is
specific to platform configuration.

The development of new models of control systems which has to
provide control and diagnostic codes, enabling the network architecture, data mapping and control and diagnostic system to be designed
and integrated in a unified and single tool, represent the main research
objective in this area.

The scientific activities and research steps consist of creating a customised process control and quality data interface system to network
stand-alone pieces of manufacturing equipment, such as PLCs, robots,
process machinery, and test stations, at all levels of the next generation responsive factory, respectively from the network of factories to
manufacturing processes.

2.6 Interdisciplinary Design of High Performance, Reliable and Adaptive Manufacturing Equipment

M
MT

Interdisciplinary design aims at supporting the mechatronics design
approach towards rapid and cost efficient and effective design and

Proactive Initiatives	**Priority**
implementation and operation of next-generation production systems. In order to achieve this, new ways of interdisciplinary system modelling for the design phase have to be developed and then exploited. The initial design phase mainly focused on insight, abstraction, cross-fertilization, domain-independence (multi-disciplinarity) aspects, has to be supported by new approaches such as structured, innovative, fast and synergistic conceptualisation and diagnostics (learn from previous mistakes). On the second design stage the following fields are of high relevance: domain-specific dynamics and control, (differential) geometry, network and graph theory, statistics and measurement, tribology, construction, etc. Tools supporting them are already available, remaining in many cases scattered, fragmented and isolated. The aim is the development of a design environment integrating the existing tools and combining them with a proper common library environment that enables quick retrieval, reuse and tracking of the corresponding design methodologies and tools. Even more essential, a methodology which supports the modelling and design decisions has to be developed. The methods should provide sufficient insight of what the design tools are actually doing and point out throughout the whole design process the need for additional expert support. One of the main benefits of this environment is not only the support of the design process, but also the contribution to the continuing education and training of the users. Such a support should also keep the designer from drifting away from the original goal (functional requirements, adaptability, life-cycle cost, etc…). Furthermore, it should take the user out of his common context and terminology and thus facilitate the communication with team members to stimulate synergy and cross-fertilisation. The integration approach is holistic, in terms of enabling flexible optimisation of multiple criteria, including default attention for sustainability in the sense of reduction of material, energy consumption, waste and noise production, addressing all aspects of both production equipment and product, both at the technical and management level. The main research focuses in this area are: (1) Development of new methods for optimisation by co-design/ establishing the right conceptual system design format (function modelling…) (2) Development of a design environment by integrating domain-specific tools, focusing on the following main research directions: integration of and creating the synergy between existing tools; simulation of holistic production systems and of machines/equipments; development of design advisory systems, used to manage complex models and the modelling process itself – decision support of transferring functional specs into	

Proactive Initiatives   **Priority**

specific domains (hardware /software); domain-independent general structures combined with domain-dependent libraries and toolboxes; standardisation and synchronisation of information which has to be exchanged between tools; distributed simulation/co-simulation

The challenge related to the design and development of the desired environment resides in the complexity of integration of heterogeneous methodologies and tools, which maintain their own business models, procedures and data locally and in the requirement of dynamical data exchange. State-of-the-art ICT has to be employed in a platform of development. Related issues to the challenges presented above are: structuring and generalising the (knowledge) content of existing tools; data management of complex mechatronic objects; combination of engineering (design) tools and planning / marketing / bookkeeping tools; interfacing between humans & simulation; Business models for modelling by industry (open format for modelling components...)

2.7 Innovative Design of Special Equipment and Tools

S
MT

The sector of tools, moulds, dies and fixtures for manufacturing is a key technology sector of European manufacturing. The definition of the requirement of these elements is under pressure in the ramp-up phases: critical time, responsibility for precision and capability, last-minute-changes, high costs. To support this critical business, it is necessary to develop and implement innovative solutions such as the following:

- Technical flexibility made by modular design, flexible automation and soft-tooling (adaptation of the software)
- Design systems (3D) with an integration of analytic methods (mechanic, thermal, electric, electronic)
- Integration of the management of objects (factory data management, simulation, virtual engineering) into the digital factory
- Distributed engineering systems
- Knowledge-based information supply
- E-tool management and remote services
- Integration of RFID and smart factory systems: ubiquitous computing, sentient computing, location systems

3 Networking in Manufacturing

3.1 Networking in Engineering

M
MT

An important factor for the successful operation of production networks is the design of the network structure and the inter-enterprise processes of it. Adequate models, methodologies, technologies and supporting infrastructures for the network design can guarantee technological, strategic and business goal alignments among business partners in a collaborative networked business environment. Furthermore, network engineering must consider the production and service capabilities of the involved companies as well as the market demands and the life-cycle aspects of the products. This engineering also includes the approach of how the product's value can be maximised collaboratively, by selecting the right partners for the joint product and service offers as well as the optimal distribution of the different adding-value steps within the network. In order to qualify the capabilities and costs of a network and different alternative designs, a network engineering methodology also has to include ways to evaluate the performance of a network at different levels of detail. This includes the definition of common key performance indicators for the different network segments as well as the ways how these indicators can be calculated. The network engineering process has, due to the complexity of the networks, to be supported by tools, allowing a detailed analysis of the network, from a static and from a dynamic viewpoint through simulation, and offering functionalities for the optimisation of network structures and processes.

The changed and still changing market demands require the frequent if not permanent design and re-design of production and logistic networks. Due to decreased product life cycles, the design of the production network has to change often. Important parameters, including production strategies such as make-to-order and make-to-stock, production and warehousing locations, mode of transport, lead times and stock levels, have to be adapted frequently. The production

Proactive Initiatives	**Priority**
networks must strive for both cost efficiency and agility to quickly adapt to the changing customer demands. Current network engineering approaches do not fulfil the new requirements: Today's network design methods for new processes and structures take too much time and effort. The time-to-market for new products is prolonged and the network design is not adjusted to the changing market demands. Consequent aspects like the market demands, company production and service capabilities as well as life cycle aspects of products have to be considered in a network engineering methodology, enabling companies to quickly assess their current network structure and identify improving areas. This engineering also includes the approach of how the product's value can be maximised collaboratively, by selecting the right partners for the joint product and service offer as well as the optimal distribution of the different adding-value steps within the network. In order to qualify the capabilities and costs of a network and different alternative designs this engineering methodology also has to include ways to evaluate the performance of a network at different levels of detail. This includes the definition of common key performance indicators for the different network segments as well as the ways, how these indicators can be calculated. The results will take the form of new network engineering methods, demonstrated and evaluated in industrial settings.	
## 3.2 Interoperable and Standardised Production Networks	M ST
Companies can be part of several production networks at the same time, thus making the planning, management and optimisation of these networks a very complex task. Research tasks are the development of organisational concepts, processes and methods for the collaborative planning, management and optimisation of production and logistic resources, including the production planning and capacity management in non-hierarchical company networks. These processes have to be standardised across industries in order to come up with the necessary speed and flexibility in the network integration. Non-hierarchical networks and the resulting decentralised planning and control processes also indicate that the supporting ICT systems for planning, scheduling and control have to be decentralised and based on distributed models and tools. The necessary seamless integration of the business processes and the supporting ICT systems require a common understanding of the exchanged information and the shared functions. Therefore, the interoperability of production networks requires a common semantic of shared information and exchanged services. With these development systems unifying the monitoring, operations	

Proactive Initiatives	**Priority**
and planning across a network, while at the same time providing the specific functionalities for the needs of a company, are possible. Therefore, new added-value logistic services, delivered by network companies, will be designed and enabled throughout a product life cycle. Widespread innovation in reverse logistics services is also expected.	
The formation and operation of production networks covers the production, distribution, after sales services, and reverse logistics. This requires a strong interoperability between the different business processes, organisational structures, but also technical solutions applied by all of the companies in these networks. The main development issues and targets are the creation of interoperable production networks in respect to reference processes, the semantics of the exchanged information and shared services as well as the application of supporting ICT infrastructures. The reference processes include the planning and execution tasks for the sourcing of materials, the production of semi-finished and finished goods, and the distribution of the finished goods to the customers. Interoperable production networks aim at enhancing the competitiveness of European manufacturing sectors by increasing the capacity of industrial SMEs to operate globally in an agile manner, in order to adapt to the rapid evolutions of existing and future markets. Deliverables will take the form of pilot implementations in industrial settings of European production networks as well as the contribution to standardisation of exchanged information and shared processes.	
### 3.3 Knowledge-Based Order Management in Networked Manufacturing	M MT
One research aspect, seen at a very large time scale, is the idea of making the orders the primary driver for adaptability, while this adaptability crosses the different levels of networked production as mentioned before. The aspect of knowledge-based product and network configurations should set the frame for the necessary adaptability of the network. While configuring the products in a knowledge-based way, incorporating the knowledge of the network's structures, processes and adaptability capabilities, the orders should be defined together with the customer, so that the product configuration, delivery performance and manufacturing costs are matched. The management of the orders should incorporate the real-time management of the network, integrating the local decision processes towards the network's wide routing of the order. Finally the ability to decide on structural changes or parameter alterations should be brought down to the	

Proactive Initiatives	Priority
local execution nodes and processes so that together with the continuous network performance evaluation self-adaptable networks will be created.	
The trend towards an international division of labour together with reduced product life cycles and the increasing importance of customised products will significantly change the engineering order management within production networks. The respective engineering departments of the network's enterprise have to collaborate closer and faster in order to meet the increase requirements, especially in respect to time-to-market and product customisation. This requires that the composition of the manufacturing and distributing network must be defined dynamically for each order. This formation of the network partners must match the order requirements with the capabilities and competencies of the manufacturing enterprises. This collaboration of different engineering disciplines requires interoperable methods supported by appropriate tools. Clear organisational structures and processes of the collaborative engineering and order management must be developed by defining activities, responsibilities as well as rights and duties of the ones involved.	
3.4 Factories Integration in Logistics Networks on Demand One important improvement through which European manufacturing companies can gain a competitive advantage is the short-term adaptability of networked production to highly dynamic changing customer demands. This covers all segments of the manufacturing network, from the product design, along the supply and distribution up to the manufacturing system supplier networks. Products have to be (re-)designed, produced and delivered to specific customer wishes as fast as possible, higher product volumes demanded in specific markets should be produced and delivered without raising the costs and the whole network should also adapt to downswings in demands very quickly in order not to focus on products still giving a substantial profit margin. By realising this adaptability, not only the network planning and control processes have to be fast and efficient, but also the single manufacturing and logistic technology has to be adaptable to customised products and product design changes, and should allow scalable manufacturing processes according to the demanded volume. Essential improvements on the operation and control level should also be realised. Here, agent control techno-logy seems to be a promising candidate for realising this operational autonomy. It is necessary to	M MT

Proactive Initiatives	Priority
develop an integrative engineering approach for the design and application of these autonomous, agile devices such as storage equipment or machinery, creating the agent-based local intelligence together with technological advances, needed for the realisation of this engineering approach. A further enhancement of the adaptability can be seen in the direct communication and co-ordination between the materials, parts and products created in manufacturing and the manufacturing and logistical equipment itself. Employing advanced ICT such as RFID allows attaching operations and control logics to the physical material flow. In the near future, traditional hierarchical and tight supply chains will have to be much more re-configurable, agile, collaborative and responsive, moving towards a self-forming supply chain and inevitably posing new and demanding challenges on its management. Research targets consist of the development of adaptive manufacturing methods for production and logistics networks. Such methods should apply modern ICT technologies and approaches for intelligent, autonomously operating machines and products. It is necessary to develop an integrative engineering approach for the application and design of these autonomous agile devices, such as storage equipment or machinery, in order to realise a high degree of agility within production and logistics networks.	
### 3.5 Networked Product/Service Engineering The networked product/service engineering research topic focuses on the segment of product engineering networks, because in this segment a lot of potential is not yet realised. In today's situation, more and more companies in a network with specific competencies are necessary to come up with the design of a new product or service. Furthermore the design process has to be accelerated to shorten time-to-market and extended to integrate the customer demands more closely. Collaborative design will embrace new methodological support and tools for understanding, tracing, and predicting usage modalities of customers throughout products life cycles, thus enabling effective product design, tailored on customer needs. New internet distributed knowledge-based CAD systems will be designed and developed. To realise networked product and service design processes and tools for the collaborative product/service design between customers, partners and suppliers, the possibility of distributing the design work across the global network has to be taken into account. To integrate and collaboratively develop the different competencies of the network companies, an integration of the knowledge resources across networks by specific processes and means has to be targeted.	S MT

| Proactive Initiatives | Priority |

3.6 Manufacturing Execution Environment for Smart Factories (Internal Networking)

L MT

Factories and manufacturing resources are permanently changing. Paradigms of the past ignored this by:

- Scheduling operations in production planning and control from month down to days and shifts
- Manufacturing execution scheduling and supervision including feed back down to hours and minutes
- Real-time control in machines down to μsec

Modern IT technologies enable the management of the factories in real-time and distribute the information of situations to all the actors in a production system. It is necessary to integrate data collection, data mining and sensors for monitoring of resources in an overall real-time architecture and present the situation permanently in the digital environment for planning, management and support of peripheral actions. This is called the "Smart Factory". Elements of the Smart Factory are:

- Wireless technology in factories
- Integration of diagnostic systems
- Real-time control and data collection for learning procedures
- Location systems for mobile objects
- Integration of Factory Data Management (FDM)
- Intelligent federation system for information supply on demand

The main objective of this research action is the development of a factory management system, based on new communication technologies and open system architectures. A federation platform for the integration of information supply has to be open for the integration of a wide spectrum of automated systems and has to support the execution by visualisation (factory cockpit) in the chains of engineering, order management and resource management. The main goal is the development of a platform for manufacturing execution, integrated in the digital factory environment of manufacturing engineering, process planning and continuous optimisation (learning elements).

3.7 Real-Time Enterprise Management

L MT

Global businesses are involved in many complex product life cycle relationships in terms of design, engineering, manufacturing, life time service support and final product environmentally-friendly disposal. This involves businesses in a vast array of different relationships with

Proactive Initiatives	**Priority**
various suppliers and customers, each having their own dynamic limitations and requirements. For organisations involved in these partnerships, there is a deep and complex requirement to manage their enterprises on a real-time basis, meeting and managing the conflicts of multiple products, multiple customers in differing supply chain positions and product life cycle stages. For true adaptability, flexibility and ability to improve productivity, these situations must be managed on a real-time basis, using data, information and knowledge, distributed throughout the supply chain and customer environment. Particular research challenges and barriers include: a drive towards true inter-operability both within and between enterprise management systems; the availability of inter-enterprise information on a real-time and intelligible basis; the ability to agree and make inter-enterprise decisions on the basis of mutual trust and benefit and finally, the ability to distribute new inter-enterprise schedules and agreements. All of these abilities can be achieved within the development of changing economic, highly adaptive and increasingly complex business process models. The development of suitable tools for enterprise and supply-chain management will support knowledge-driven outsourcing business models on a global scale.	

4 Knowledge-Based Manufacturing Engineering

4.1.1 Renewing Industrial Engineering

S
MT

Manufacturing engineering is the key technology to implement innovations and to design products, services, processes and manufacturing systems. The implementation process requires the employment of efficient tools, based on the state-of-art knowledge, expertise and best practices in manufacturing engineering.

Manufacturing enterprises, called factories, have to rethink their organisational structures and basic activities to accommodate the changes foreseen in manufacturing processes. Manufacturing engineering addresses simultaneously all interrelated aspects of a product life cycle from design to recycling and disposal.

The area of manufacturing engineering is the centre of manufacturing development. It is embedded in networks of product engineering, material and component suppliers, manufacturing suppliers and customers. Manufacturing engineering processes take place in the manufacturing system.

Manufacturing engineering is a holistic approach including the engineering of the factory structure, the development of the organisation, the design engineering, the process engineering and the development of the required tools and application systems.

At all levels, e.g. manufacturing network, segment or system, machine or equipment, subsystems and processes, the factory and its manufacturing processes can be defined in their 'current' and/or 'future' states, under the so-called 'digital' and respectively 'virtual' representations. This relates to the employed models, methods and digital tools or simulation applications and systems used to represent the static or the dynamic states.

4.1.2 Digital Manufacturing for Rapid Design and Virtual Prototyping of Factories on Demand

S/M
MT

Consumer needs and expectations of the future will require a continuously and rapidly evolving production framework: thus production

Proactive Initiatives   **Priority**

systems, from small to large scales and integrated factories, shall be conceived and set up in more and more shorter times. This will require a conception and development of new methodologies and innovative tools, which enable and support the rapid design and prototyping of the entire production system. The creation of a holistic, up-gradable, scalable virtual factory can foster high cost savings in the implementation of new manufacturing facilities, thanks to the effective representation of buildings, resources, process and products. Decision makers and designers can benefit from the closer integration of product, process and plant developments through advanced modelling and simulations.

- Development of a virtual factory framework, focusing on a development of a reference (standard) factory data model, generic architecture for collaborative virtual factories and products through integration of heterogeneous models, methodologies, technologies and corresponding tools of digital and virtual factories;
- Employment of virtual/augmented and mixed reality technologies and tools for enhancing the factory experience and the immersion and presence of humans in the environment of virtual factories;
- Development of technologies and tools for factory, product, process modelling, simulation and virtual prototyping;
- Support for the integrated process/product engineering and implementation of production processes simulator architectures for the automation systems development and configuration;
- Integration of digital manufacturing technologies (industrial process simulation).

The envisioned main results of these research activities represent a complete detailed framework for the virtual factory and tools for the quick, reliable and optimised creation of knowledge-based manufacturing systems and factories. They should enable collaborative, interdisciplinary and multicultural design/analysis and optimisation of processes to be executed effectively and efficiently in global virtual company networks. The required tools should consist of software, using intelligent databases and data analysis and presentation methods, complemented by models, processes and guidelines enabling their usage. Their capability needs to be proven through successful employments in European manufacturing companies, resulting in significant measurable improvements of business success indicators like time-to-market, customer satisfaction, market share and revenue as well as in improved soft factors like working climate, quality of life, environmental protection and innovativeness.

Proactive Initiatives | **Priority**

4.2 Configuration Systems: Customisation of Products and Services to the Market Requirements	M ST

4.2 Configuration Systems: Customisation of Products and Services to the Market Requirements

Configuration systems are key tools in supporting the enterprises to quickly bring out innovative and profitable products to the market. The current systems for product services customisation allow customers to specify their requirements by selecting and configuring products and the related services. Customers' individual requirements beyond product configuration and adaptive configuration of complex items (products or services) cannot be fulfilled in these systems. However, such systems, typically designed for expert users, are too technical for the average customer. The integration among the product customisation system and other application systems of the enterprise is rarely considered. Systems enabling enterprises in order to meet the customers' individual requirements more effectively, by providing more customisation approaches, represent one of the main requirements related to the design and development of the new generation of modern manufacturing systems. Through system integration for continuous manufacturing management, the system as a whole eases enterprises to optimise their product and service development processes and supply chains according to the customers' requirements in a systematic way. Several weaknesses of the current systems have to be overcome:

- The current systems are designed for one typical user class, not approaching the fact that users differ in needs, knowledge about the product details, and expertise.
- Many configurators are product-oriented in their communication process, ignoring the needs of large user groups, e.g. goal oriented customers, who cannot deal with, are not interested in product details.

Design and implementation of user-adaptive configuration systems for products and services towards the enhancement of the existing customer relationship management systems (CRM) represent the main research activities that have to be performed.

The modern user-adaptive configuration systems have to cope with the above mentioned challenges through:

- Enhancement of the usability of already existing configuration systems by extending them with user-adaptive interfaces, supporting and guiding the user through the configuration process in a personalised way

Proactive Initiatives	Priority
• Development of new models of representation of knowledge about products and services, having as a result the so-called customer-driven product and service generic model • Development of new techniques to support the adaptive configuration of items, which is essential to comply when purchasing a complex product, or registering for complex services, through a user-driven management of the product and service model, the personalisation of the interaction, and user-adaptive explanation of conflicting requirements is identified in the customisation process • Development of new modes of integrating the user interaction with configuration, through intelligent user interfaces mediating the user and the configuration system. The integration of the envisioned user-adaptive configuration systems for products and services with the CRM systems improves the overall communication with the user and supports the satisfaction of users' individual needs at the cost of mass-production.	
## 4.3 Tolerance Systems for Micro and Nano-Scaled Products	S MT
For the normal use, industrial tolerance systems are standardised down to μm. In the dimensions of micro and nanometer, they have lacks because of extremely influencing factors of capable measurement (roughness, form, position) and influencing factors of the environment (temperature, contamination by particles etc.). The tolerance systems have to be scaled down to support the reproducibility of parts and components in combination with measurement procedures and technologies. Problems of calibration, management of measurement and of high precision technologies have to be solved by industry. Tolerance systems have to be integrated into the design and quality management, especially under the aspects of micro and nano manufacturing, aiming at increasing the reliability of micro- and nano manufacturing as a base for future standards and the design of reproducible parts and components.	
## 4.4 Knowledge-Based Process Engineering	
### 4.4.1 Hybrid Systems	S MT
Hybrid systems are characterised through a mixed and changeable degree of automated and human work. Human work is essential for changing operations like in a series production with high numbers of	

Proactive Initiatives	Priority
variants and customised orders. Process planning, supporting the preparing and optimising of manufacturing operations, has to execute the work under strong time pressure and with high accuracy. Defects in process data and process parameters, programmes for automated operations, are creating losses and defects in the shops. They need tools for efficient planning in the chain between engineering and execution, with a link to the shop real situation. A possible solution for the above mentioned challenges represents the embedding of knowledge in the process planning by means of elementary standards of work (global process standards), experience-based and cognitive learning, data, knowledge and best practice-based integration in real-time manufacturing execution systems and real-time resource management. The implementation of vision systems for processes in distributed (networked) manufacturing and for capability of resources have a main relevance for this topic. In recent years, the developments of vision systems have concentrated on solutions that detect defects earlier in the process or prevent them from being created at all. As an example, so called vision labs have been deve-loped, designed to automatically sample sets of containers from the production line and perform a series of highly accurate tests. These measurements are monitored with special software for control variations, where actions can be taken to prevent the problem drifting to an out-of-specification state. Thus, quality can be maintained or improved, efficiencies increased, and costs minimised. Several solutions and tools should support the objectives' achieving: knowledge support of the so-called Intranet federation platform, cockpit application for planning and execution in partly autonomous socio-technical environments of manufacturing.	
4.4.2 Process Planning in a Customised Production Regarding the increasing demands concerning flexibility and cost reduction in car body manufacturing, flexible forming systems which can be used for the realisation of complete car body parts, offer the chance to achieve considerable economic effects. These systems can be used for the realisation of different parts of the same part family (e. g. doors, bonnets ...) or for the manufacturing of complete parts of different part families. Using these systems, it is possible to form the main part geometry as well as special designs or functions defining features. The main component of such a manufacturing system can be represented by a forming device, which consists of a forming tool (in a press) or a 'self-driven' forming tool, respectively. The forming tool itself can be based on several components, such as 'multi-useable' modules or part-related segments (e. g. for corner	M MT

Proactive Initiatives	**Priority**
areas). Beside the main forming steps, other operations can also be integrated (e. g. calibrating, punching, trimming, joining …). An important precondition for the realisation and use of such manufacturing systems is the development of a method in order to identify the process steps for the realisation of the special part geometry. Following know-ledge-based aspects have to be included: • A scanning system of the part geometry (CAD data) • An automatic part classification • An automatic identification of the main geometry and special geometric features • A determination of forming processes and required tool components To guarantee the feasibility, automatic FE simulation and optimisation loops (if required) should be also implemented. Deliverables include (i) methods for the development, realisation, configuration and reconfiguration of flexible forming systems, (ii) the automated coupling of the process planning and forming device configuration with FE simulation and (iii) a prototype flexible forming system. ## 4.5 Digital Libraries and Contents for Engineering and Manufacturing In many fields of the engineering area, such as electronic engineering or mechanical design, the new designed and developed products are mainly based and then defined of already existing components. In order to increase the efficiency and the quality of the design process, the digital libraries have to feature a high level of availability regarding stored information on the already existing components, and of exchangeability or remoteability access through networks. In the last years, several requirements emerged, regarding parts and components catalogues. With the development of digital mock-ups in a number of industries, the need is to fully access digital representations of all the parts intended to be used in new products. A solution for this typical data modelling problem represents the PLIB (ISO) standard, which uses EXPRESS information modelling language. The development of electronic versions of paper catalogues and their distribution over the Internet, identified as typical document structuring problem, has as a possible solution by using the XML tagging technology. Since the PLIB approach is not suitable for the purposes of browsing, presenting and understanding this information, the new XML-based exchanging technologies seem to be appropriate, but not enough reliable to be	 S MT

Proactive Initiatives	**Priority**
used by applications that should exploit the meaning of the structured information. The conversion digital library information in catalogue format, the publishing of this information in the Internet and then the integration into the factory collaborative information environment, represent a challenge which will be faced in the next years by fitting the needs of engineering and manufacturing with the progress in the advancement of digital libraries ICT research field. The research aims at defining the digital library services as a key component of factory digital infrastructures, allowing content and knowledge to be produced, stored, managed, personalised, transmitted, preserved and used reliably, efficiently, at low costs and according to widely accepted rules, standards and protocols. Several scientific goals are envisioned: • Ensuring the long term accessibility and usability of the content of the digital factory, respectively parts and components, which have to be available in digital form through digital libraries, • Development of new and more effective technologies for intelligent content creation and management, and for supporting the capture of knowledge, its sharing and reuse, • Development of new methods for supporting people and organisations to find new ways to acquire and exploit knowledge, and thereby learn. Design and implementation of the so.called authoring environments for engineering and manufacturing, aiming at supporting the design activities, based on new forms of interactive and expressive content using and motivating the multimodal experimentation and exploration of the design space. These design environments for engineering and manufacturing will facilitate the content, sharing automatic tagging (XML DTD) of existing multimedia content of parts and components. This content will be stored by using open standards, as annotated output in scalable repositories, enhanced with integrated indexing and search capabilities.	

5 Technologies for Future Products

5.1 Integrated Technology Management in Design-Intensive Product Environments

S
ST

The European manufacturing industry is shifting its focus towards design-intensive knowledge-based products with integrated services via an ever-increasing involvement of leading-edge technologies. This shift affects all industries, but the most revolutionary rethinking is fuelled by the fierce competition in mass production segments that are close to the consumer and rely on fast changing technological environments. European economy can only react by forming inter-meshed trans-sectorial technology networks of SMEs and large global companies that are leading innovators in their areas and will enable strong technology platforms.

In this environment, technology management plays a decisive role because it can empower European high-tech companies to identify, analyse and implement those technology platforms that are vital to their success.

European technology companies have to become the leader in technology management if they want to stay leading in technologies. This will be made possible by the results of the projects under this topic. Methodologies will be developed to integrate technology management efficiently into European management procedures. A trans-European roll-out of increasing technology management activities has to be enabled by the assessment of the best practices and a strong dissemination background, which will be tailored towards the specific European needs.

5.2 Technology Monitoring and Scanning

S
ST

In a technology-driven competitive environment, management of technology has to steer the build-up and the usage of technological competences in a company. Technology monitoring and scanning are

Proactive Initiatives	Priority
the first of the four central steps of a state-of-the-art technology management process. A fully functional technology monitoring has to identify candidate technologies for the following steps of assessment, planning and the usage of technologies. Effective technology monitoring assures that attractive new technologies are identified early, their development can be predicted and expected discontinuities in the development of technologies are detected faster than the competition in order to be able to react to these insights. Technology monitoring itself consists of the determination of information needs, of obtaining and analysing information and of communication. In recent years, technology monitoring has, especially due to the ever rising efforts that are undertaken in cooperation with other companies or bought externally, become an even more important and complex task for companies. Technological know-how is bought externally in the form of components or technologies. Because of this reason the internal view of the innovation and technology development processes has to be expanded by the external perspective, which consists of external acquisition and the usage of technologies. As important a process as technology monitoring and scanning is for the mid- and long-term success of any knowledge-based enterprise, as hard it is to implement it efficiently, especially for SMEs who lack the international resources of global players. The central challenge for SMEs is to, despite of their limited resources, still cover a broad range of interests with a thorough assessment. In most cases this can only be realised through networked co-operations. Projects answering to this call should focus on the development of toolboxes that enable European SMEs to monitor and scan technologies with a networking approach. Based on this, the specific needs of knowledge-based European SMEs need to be assessed and new methods, tailored towards these needs, have to be developed. The industrial involvement should be strong and results need to be validated through quantifiable positive effects in industrial trial applications.	
## 5.3 Next-Generation High-Adding-Value Products Science-based high-adding-value products are a key result to be achieved for moving the European manufacturing sector towards a new competitive advantage on the global scale. Such an RTD activity needs a strong exploitation of world-leading developments in enabling technologies such as new materials, nano-, bio-, info- and cognitive technologies. Next generation HAV products for the final consumer have to be 100% personalised, comfortable, safe, healthy, and eco-sustainable.	M MT

Proactive Initiatives

Therefore, the following major RTD sub-topics have to be addressed:

- Introducing innovative sensors, actuators and embedded cognitive technologies for active products, supplying functionalities and services for comfort, health and safeness of the consumer;
- Introducing bio, micro- and nano-components, as well as intelligent and multifunctional materials, for self-adaptive and eco-sustainable products.

Main development issues and targets, and deliverables are:

- Methods and tools for forecasting consumer attitudes and needs based on social and cultural aspects to conceive disruptive new products-services, anticipating the market dynamics;
- Knowledge-based collaborative environments for the design of next generation products, integrating new materials, nano-, bio-, info- and cognitive technologies;
- New manufacturing processes for next generation consumer oriented science-based products.

RTD activities have to be developed with reference to relevant manufacturing sectors as benchmarks with reference to:

- Traditional industry (e.g. textile, wood and leather products);
- Mass production (e.g. automotive and white sector);
- Specialised suppliers (e.g. aerospace, machine tools);

In order to shift such manufacturing sectors towards more science based HVA solutions.

5.4   High-Added-Value Product Design and Virtual Prototyping

S/M
MT

A market's success of new products or services is largely determined by decisions taken during the design phase. An interdisciplinary and intercultural design team needs the ability to anticipate the future preferences of the customer in order to be able to develop attractive products. Successful development projects focus on the integration of customer influences in the design and development process and the related demands for manufacturing processes and thus enable intelligent customer-driven innovations. The need for a deeper integration of customer preferences from the timing, quality and product capabilities perspective creates entirely new challenges both in the execution and the management of development projects. These challenges are of a threefold nature:

Proactive Initiatives   **Priority**

- Commercial challenges are ranging from distributed supply chains and networked business models to intelligent new selling and payment methods
- Social aspects include for example the integration of multi-site and multinational/multicultural development teams as well as culture specific customer preferences
- Technical issues are to be found in the realisation of radical new product features and intelligent production systems, as well as in responding to strengthen environmental and socio-economical questions.

For the team members of development or management areas it is, due to the ever rising complexity of their tasks, increasingly difficult to keep up with the pace that is required in order for the European industry not to lose its already weakened position in global manufacturing. Their ability to cope with the above mentioned challenges needs to be strengthened through common European efforts, aiming at enabling European companies' product development teams to respond quickly, efficiently and effectively. For this, they need to be put into possession of tools that are tailored and fine-tuned for the specific European characteristics of product development, considering the European approach to industrial design. These tools need to enhance and enable the usage of globally recognised European capabilities and must therefore be consisting of software tools as well as of radical new methodologies, business processes and best practice guidelines.

The main issues and targets for the development of these tools are:

- Enabling the development of intelligent products and services with multiple and adaptable capabilities through knowledge based design for the integration of multiple functional-adaptive and self-optimising systems;
- Collaborative and multidisciplinary product design in virtual global company networks, using optimised standards, data exchange formats and product description features;
- Virtual mock-up, including enhanced virtual production design (including transfer to ramp-up and operation through diagnosis, simulation, data analysis) and prototyping with improved human-machine interaction;
- Life cycle value management, for example implementing customer value optimisation through life cost modelling or user centred business models and using reliable prediction methods;
- Rapid customer driven product/service development, enabling customer-designer-manufacturing interaction and mass customisation;

Proactive Initiatives	Priority

- The central issue is the realisation of new methods for reliably predicting costs and greatly lowering time to market while highly improving product and service quality/ features and customer-orientation. By doing so, the social aspects of innovation and the challenges of the management of cultural aspects in cross-company international networks, need to be respected;
- Advanced (research integrating) design methodologies and tools to reduce time to market of research.

All these issues and challenges have to be overcome through the development of tools for the quick, reliable and optimised creation of knowledge-based products and services, enabling collaborative, interdisciplinary and multicultural design processes to be executed effectively and efficiently in global virtual company networks. The required tools should consist of software, using intelligent databases and data analysis and presentation methods, complemented by models, processes and guidelines enabling their usage. Their capability needs to be proven through successful employment in European manufacturing companies, resulting in significant measurable improvements of business success indicators like time-to-market, customer satisfaction, market share and revenue as well as in improved soft factors like working climate, quality of life, environmental protection and innovativeness.

5.4.1 Extended Product Services through Integration of Product Life-Cycle-Knowledge into the Products Themselves

S
MT

Information, knowledge and documentation about each product are part of the delivery and are important for product oriented service operations like maintenance or training. The expenditure of documentation increases with the complexity and customising of new and high-value adding products. Product documentations are generated during the engineering process and are adapted by manufacturing and service. They can be combined with diagnostic systems to have a permanent state-of-the-art and the on-line availability (24 h/day within few minutes). Such systems represent prerequisites for global sales and services especially in the machine industries. Together with e-maintenance activities a European way for service support can be implemented. Main challenges on this way are the reliability and security aspects as well as economic aspects of global services. New and enabling technologies for the efficient generation of information and documentation by manufacturers and the integration in the life cycle management represent a fundamental aspect for efficient growth in the 'after sales' business. In the development of life cycle information and documentation systems with efficient data management, the integration of Internet technologies and applications should take into account the complexity of IP and security aspects.

Proactive Initiatives	Priority
5.4.2 Condition Prognostic Capabilities for Improved Reliability and Performance The research should focus on extending the condition monitoring system with prognostic capabilities. To this end, an explicit (physical) and/or implicit (e.g. neural network) model of the degradation behaviour of the machine components over time, in function of the operation conditions, is required. The appropriate format and the methodology to obtain these degeneration models in a cost-efficient way should be established. In particular, different learning approaches for individual machines or classes of machines could be a valuable contribution to this aim.	S MT
5.4.3 Revenue Optimisation through Condition Monitoring and Prognostics The research should focus on using prognostic capabilities to maximise the revenue of production machines. 　　In particular, the short to mid-term operation modes of the machines and the maintenance schedule can be optimised for productivity, maintenance costs, energy consumption, etc. Various optimisation methods should be compared to determine the optimal way for obtaining the maximum added value from the production equipment. A gradual migration path from operator advisory systems to full self-optimisation should be the intent.	S MT
5.5 Sustainable Life Cycle Management of Factories and Products The modern view on manufacturing engineering resides in incorporating the life cycle paradigm into the factory as a whole, its corresponding products, manufacturing processes and technologies. The idea of 'product life cycle' is essential for the path to sustainability by expanding the focus from the production site to the whole factory and product life cycle. The main goal of factory and product life cycle thinking is to reduce resource use and to improve the technical and social performance, in various stages of a factory and product's life. Life cycle management is the application of life cycle thinking and models to modern manufacturing engineering practice, with the aim to manage the total and comprehensive life cycle of the factory and its products and manufacturing processes and services towards more sustainable consumption and production. 　　Life cycle management is about systematic integration of the product sustainability into the manufacturing strategy and planning, product design and development decision making and communication and collaborating applications. By implementing the life cycle management	M MT

Proactive Initiatives   **Priority**

capability, considerable benefits, such as faster time to market, lower costs, reduction of rework and rejection dates and more component and technology reuse are achieved. This approach gives the image of a three-dimensional life cycle space for factory, products, and manufacturing processes. Each of these entities has its own life cycle, consisting of specific phases.

Each factory follows a life cycle from its initial concept in the mind of an entrepreneur to the ecological dismantling, through a series of stages or phases. Despite the identified and recognised phases: design and planning, construction, operation and maintenance, refurbishment or obsolence and end-of-life phase or dismantling, this work focuses on the first phase, design and planning the factory. In this phase, in great interdependence with the life cycle of products and used technologies, the factory processes and its production facilities are planned.

The central part the overlapping of the factory operation and maintenance and the manufacturing of products in the so-called production phase, represents the crucial and at the same time critical point, called 'crossing-life cycles point'. Here, virtual products and factories become reality. The real product is built into the real factory. Then, the manufacturing processes are implemented by using the most suitable technologies. At this point, all the already performed engineering activities and efforts are to be proved and verified. In this phase, the real factory has to be highly transformable in order to quickly respond to the changes occurring in the product world: frequent product launches, increased product complexity as a consequence of using advanced and emerging technologies, e.g. the fast development of micro and nano electronics, increased micro computerisation, and new materials development.

The crossing life cycles point shows the results of the preceding phases concerning the manufacturing of products under optimum conditions (time, quality, costs). The point not only highlights the efficiency and effectiveness of the used models, methods, technologies and tools for planning and designing products, processes and factories in digital and virtual world, but also the appropriateness of using them. The main advantage resulting from this approach is the transformability and changeability of the factory structures throughout their whole life, according to the manufactured products, the corresponding manufacturing processes, and the technologies used under economical conditions. Thus, in the operation phase the factory is already prepared to react to a change regarding a new release of a traditional product or a new product, a newly implemented state-of-the-art manufacturing process or the use of an innovative technology. These foreseen and possible changes have already been taken into consideration in the planning phase. Then, the factory is able to respond adequately and to adapt itself to these changes and turbulences in order to remain

Proactive Initiatives	Priority

competitive. The information gathered in the production phase represents a valuable input for continuous re-planning and adapting.

5.5.1 Unified and Sustainable Life Cycle Challenges and Risks

S
MT

An orchestration or harmonisation of the specific life phases of product, manufacturing processes and technologies with the planning phase of the factory represents a great challenge. This approach is called unified and sustainable life cycle management.

There is a risk associated with the things in the world, which have a life cycle themselves, as in a case of a factory. The manufactured products, the corresponding manufacturing pro-cesses and the technologies used, all these subordinated factory entities, have their own life cycles. Each life cycle can be represented, at the end, as independent software application; therefore, a software technology infrastructure has to be formulated to allow for the seamless linkage and integration of software application and systems, representing various life cycle aspects. Because phases of these life cycles tend to be independent of each other, the current challenges and then the research efforts have to be coordinated towards integrated and unified life cycle paradigm. This unified life cycle paradigm builds upon current technologies and is backwardly compatible while embracing future emerging technologies. Only when two of these life cycles coincide and one affects the other there is connectivity and a transfer of information at the interface. The current research approaches have to identify 1) linkage points (i.e., portals) between life cycles, 2) the type, and form of data passing between life cycles, and 3) conditions when life cycles interact and communicate. This is expected to be overcome by developing and integrating new technologies and tools, e.g. information and communication technologies (ICT), digital manufacturing technologies, collaboration models and tools, used to trace factories, products, processes and technologies over their life cycle from engineering to the end of their lives. Several strategies to support the required orchestration have to be mentioned: applying the simultaneous engineering for bridging the product design and process planning, and the development of suitable strategies for R&D in order to link the product planning and development and factory investment and engineering. The last can be achieved through the development of advanced and innovative manufacturing technologies.

The envisioned solution for minimising all risks and losses related to the crossing life cycles point, is the development of an environment for factory life cycle, by collaboratively integrating the latest technologies and tools used to follow the factories and their products along their life cycles. The vision of this work represents the 'transformable and adaptable factory' which has to react quickly and appropriately to the internal and external turbulences, by using new collaboration and integration models, methods and procedures along the value chain.

6 High-Performance Manufacturing Technologies

Manufacturing is expected to keep on increasing its demands in terms of quality, reliability and productivity. This tendency has been clear in the last 5-10 years, induced by the shortening of the time-to-market processes, the faster changes in the products presented in the different markets, and the increasing competence of the low cost countries.

The big manufactures (consequently, machine consumers) are producing more accurate parts in lower times, with a shorter lifetime on the market, and under more environmentally friendly parameters. These requirements are directly translated to the machine manufacturers, that are more and more being transformed into "manufacturing solutions generators", more than simply machine suppliers.

This new requirement can only be satisfied by the development of manufacturing systems, based on technologically advanced solutions, usually, integrating several high-value technologies into each solution.

The most significant aspects that should be covered by these new machines and processes will be: (1) technologies (machines and processes), that provide more efficient and productive outputs by high volume, high speed and capability of Processes. (2) technologies (machines and processes), that overcome the current frontiers in terms of accuracy, providing smooth and even super finishing conditions. (3) technologies (machines and processes), with special mechanical and control characteristics and configurations in order to provide drastic improvements in process dynamics, increasing the productivity in the same level. (4) technologies (machines and processes), that require less shop floor space, by means of the reduction of peripherals, optimisation of machining cycles and process planning.

The main aspects that will be the subject of investigations under the scope of this topic should be: (1) drastic improvement of the conventional manufacturing processes, by means of new technological approaches, based on new strategies, tools and machine attributes.

Proactive Initiatives	Priority
(2) development of alternate processes that substitute the conventional ones or combine with them, provide new productive, economic and ecological ratios. (4) development of new machine (equipment) concepts, based on new materials (including nano and smart), new architectures and new control possibilities.	

6.1 Efficiency of Resources Energy and Material

### 6.1.1 Total Energy Management	S ST
Different systems of energy supply are used in manufacturing: process energy (heat), energy for driving machines and systems, energy for working environments, light, special operations and regeneration of media and material etc. Seen from the point of the cost of energy in manufacturing and impacts on the environment, a general initiative has to be started, which initiates a reduction of energy consumption in manufacturing. The scope of this action is a Total Energy Management System (TEMS), achieved through the following main research activities: • Development of an energy management system, similar to quality management, which includes the life cycle from engineering to disposal and recycling, • methodologies for preventive energy reduction: design to decrease energy, assessment/evaluation of energy • Innovations to reduce the energy consumption of machines, energy supply systems, infrastructure, logistics and buildings	
### 6.1.2 Technologies for Energy Efficiency, Consumption and Scavenging	S ST
Optimised utilisation of energy streams with a low energy level is a very promising approach to reduce energy consumption and to increase energy efficiency in production processes. The detailed knowledge and analysis of the production process is the prerequisite to open up energy saving potentials in manufacturing industries. The objective will be to overcome existing process limitations by developing new production processes which integrate innovative energy efficient technologies. Optimised utilisation of energy streams with low energy levels especially includes the application of innovative approaches and technologies for the utilisation of waste heat.	
### 6.1.3 Clean Manufacturing Processes	S ST
In the automotive and machinery industry the structures in metal or polymers are getting smaller. Particles in these structures are reducing	

Proactive Initiatives	Priority
reliability and quality. The objective is to develop clean manufacturing processes to avoid particle contamination in products. The focus will lie on the protection of the clean environment for conventional production processes like moulding and shape cutting. The environment assures the cleanliness with airstreams, monitoring systems, filter systems, inline cleaning and other principals known from semiconductor industries. To ensure the clean manufacturing environment advanced technologies for particle detection, cleaning technologies, sensor technologies, contamination suited product design and the cleanliness of materials have to be improved. As a concrete need, a new approach to avoid contamination in conventional industries has to be developed. A clear strategic contribution to establish a European high value added industry is expected. New, cost efficient production methods will improve the quality of products in high market value segments in industries such as automotive and machinery industries. ## 6.2 Technical Solutions and Components ### 6.2.1 Optimal Energy Consumption by Flexible Self-optimising Drive Concepts The main objective is the flexible adaptation of electric-fluidic energy resources for high performance drives both to production system and to process needs to overcome traditional efficiency limitations of local energy sources (hydraulic and distributed pneumatic power stations) by concepts of generating energy on demand and feed-forward strategies. A higher performance (speed, acceleration) is usually limited by a higher installed electrical power. But energy cost share in product prices increased from 10% to 20% in the last years and energy prices still increase on the market. Research should focus on flexible drive concepts for altering demands of process conditions. Several such types of drive should be mentioned: a wide range of volume flow or velocity, mass or acceleration and jerk-free movements with regeneration of accumulated potential and kinetic energy (e.g. servo press or direct drive technology); combinations of direct electrical drives with rechargeable batteries providing highly reliable and safe energy and drive concepts. Local energy sources cooperate with production systems on the basis of new mechatronic model-based or knowledge-based motion control and real-time sensor applications to realise forward energy planning	 S MT

Proactive Initiatives   **Priority**

demands; co-operation of multiple main or servo drives in motion control and energy regeneration are highlighting the same relevance, as well.

New generations of adaptive production systems with increased drive or process performance of adaptive production systems by 20% and decreased local power consumption by 25% represent one of the valuable outcomes in this area. Reducing the waste of European energy resources in local industrial energy generation and motion consumption of production systems can be achieved as well. In the wake of drastically dwindling energy and material resources the need for such systems becomes increasingly evident in the field of new and old industrial automation systems.

6.2.2 Self-optimising Electric-Fluidic Energy Sources for Optimal Energy Consumption

S
MT

Innovative technologies for temperature control, power generation and storage in production processes open up a big energy saving potential. They are essential to adapt to manufacturing processes in order to become more and more energy efficient. Taking into account the worldwide increasing energy demands and the increasing greenhouse gas emissions, it is obvious to focus on energy efficiency. Of central importance is the integration of their application into the whole energy concept, including all types of energy, used in the production process.

The objective will be the production and the application of innovative energy generating and storing technologies, such as polygeneration plants, organic rankine cycles (ORC), or phase-change materials as energy carriers in production processes and the use of refuse derived fuels (RDF, e.g. from industrial waste streams). The expected projects should perform research integrating cross-cutting technologies for the energy supply of heat, steam, cold and power, aiming at the development and application of innovative temperature control, power generation and storage technologies. Thus, a broad range of competences is needed to organise and accomplish the accordant projects. The expected benefits especially for manufacturing and production processes are (i) substantial energy savings in industrial processes, (ii) technologies for the closed loop on the second and third level of energy streams, (iii) new energy efficient production processes, (iv) reduction of greenhouse gas emissions, (v) protection of fossil resources.

6.3 From Atom to Parts and Components

6.3.1 New Material Functionalities Induced by Manufacturing Processes

S
ST

The interaction of manufacturing processes and the product's materials has a considerable influence on the properties of these materials.

Proactive Initiatives   **Priority**

Sometimes such effects are used to realise desirable functions of a material, but sometimes it is just an unnoticed side effect. A better understanding of such interactions provides the knowledge needed for completely new ways to realise properties of materials or material assemblies such as sandwich structures.

The careful planning of the process-material interactions, their sequence and the control of these processes enable a far better exploitation of the material's functional potentials. Appropriate manufacturing techniques will by far enlarge the processing windows and yield materials that are not used industrially today. Additionally a carefully controlled process-material interaction can produce customer specific product variants that are not economically feasible today. It will even enable new strategies for the modularisation of products and production concepts. New functionalities will be achieved by innovative manufacturing technologies for joining and integrating materials and components that do not fit together today. Advantages for new and innovative products lie in their adaptation to customer specific use and environments.

The activities planned to be performed in this research area have to focus on:

- Manufacturing of more individual, user adapted products at lower costs by using unexploited material,
- New concepts of built-to-order products with decreased complexity compared to the state-of-the-art.

6.3.2 Process Planning for Multi Materials and Functional Material Manufacturing

S
MT

The need for light weight constructions as well as integrated functions and miniaturisation leads to an increasing use of multi materials and functional materials in several fields of application.

In the field of light weight constructions the high strength materials are used for low weight and high rigidness. To also achieve appropriate damping behaviour, multi materials as combinations of metal sheets and plastic sheets, are applicable. Also functional combinations of 3D shaped parts, made of different materials, can contribute to a lowering of the weight of constructions, especially in vehicle constructions.

Also the increasing functional integration as the integration of mechanical, electrical, fluidic and other elements into one part, during an early stage of machining, is essential for short and reliable processes.

For this reason there is a need for the planning of design and machining of functional integrated parts. This will include the design of

Proactive Initiatives	**Priority**
multi materials and multi material parts, the investigation of machining processes for the assembly of materials and shaping of multi material parts and the assembly of shaped multi material parts. In addition, the functionalisation of mono materials by an integration of active and elements is necessary.	
6.3.3 Manufacturing of Engineered Materials In the area of materials engineering, one of the main objectives is the reliable large scale production of engineered materials. The main challenge represents the strong interaction between the material properties of the different material components in engineered (composite) materials. So it is necessary to implement a detailed on-line measurement system to identify quality parameters at the several steps in the process chain. New algorithms for online process controls allow a delicate design of new production processes. An online-documentation of quality parameters is necessary to guarantee the trace of relevant data of product and production technologies. The research should be focused on the development of new and innovative technologies aiming at increasing the reliability and reproducibility of the so called smart composites. A smart composite is the combination of a light weight material like fibre composites or light metal alloys with a sensor and actuator material like piezo-ceramic or shape memory alloys. So the inherent sensor effect can be used for a control of the production process and a health monitoring during the process chain.	M MT
6.3.4 Engineering of Integrated Materials The increasing functional integration such as the integration of mechanical, electrical, fluidic and other elements into one part during an early stage of machining is essential for short and reliable processes. New functionalities of mono materials by integration of active and passive elements are necessary. Process engineering and system engineering of integrated materials need innovative tools and methods for combining basic and adaptive materials in energy, information and material flow conditions. Contact between basic and adaptive materials shall provide reliable initial pressure and resist mechanical stress in manufacturing process, training and applications without delaminating or cracking. Today's usable adaptive materials for sensors and actuators are controlled by external electrical current, temperature, and forces. Electrical conductors, isolators and mechanical interfaces with microstructures in large dimensions have to combine in reliable production processes into basic materials. New material integrated controllers are	M MT

needed, made of non-silicon materials which can be programmed for controlling positions or force in closed loops. Polytronics offer printable electronics on polymers but do not resist high temperature or mechanical stress. Integrated materials demand new chemical layer processing (without liquids), which do not destroy previous integrating process steps. Mass production processes demand a higher level of automation and cost effectiveness in the whole process chains. New concepts and techniques for product integrated micro quality management of basic and adaptive materials are needed. Concepts for environmental friendly regeneration of integrated materials are expected.

6.3.5 Manufacturing of Advanced Materials and Functional Surfaces

L
MT

The need for light weight constructions as well as the use of high strength materials and miniaturisation leads towards an increasing use of multi materials in several fields of application. The trend of miniaturisation leads to 3D surface elements with micro cavities (hole, channel) and increases the influence of micro/nano forces, not only between surfaces, but also between elements and the environment. Advanced materials are characterised by a significant change in mechanical stiffness, hardness or by a combined electrical conductor and isolator or chemical activeness and resistance. As they consist of different chemical elements or slides with different behaviours (like bimetals), the material properties have "to be trained" in production conditions for a life long application. New adhesive forces between different material elements have to resist external application forces (like cutting, forming, joining) without delamination or waste. Tools are needed for engineering and manufacturing of determined advanced material behaviour.

Functional surfaces are characterised by a significant changed parameter as surface structure or roughness for friction forces or optical reflection, in micro adhesive forces or metallic corrosion of coupled surfaces, for clean production and so on. Knowledge of micro/nano material behaviour of known macro materials is necessary to solve new nano technologies in cost effective applications. The trend of replication technologies with micro and nano structures towards the highest precision manufacturing systems limits today's field of application.

6.3.6 Manufacturing of Graded Materials

M
MT

To improve the performance of single parts and in general of mechanical components it is more than ever necessary to implement more

Proactive Initiatives	Priority
and highly complex functionalities. One way to fulfil such requirements is the usage of multiple materials inside of one part and the development of new technologies to be able to realise graded material properties. The controlled integration into the product's design development and application of graded properties is representing one big step forward into complete new design and process technologies, which enable the production of until today unknown structures, part geometries and part properties. This new challenge is not only focused on new technologies to realise those properties especially the graded one, but on the whole product development process. New CAD-software and tools are necessary to realise the design of graded materials and the implementation of material data inside of today's 3D modelling tools. In terms of international data exchange and the global usage of a standard format, a new and common format, where geometrical information and the dedicated material properties are described, must be defined. Most of today's known technologies are not flexible enough to realise controlled and predefined properties inside of one part. New technologies based on additive techniques must be developed to enable controlled build ups of different materials and graded properties inside of one single process. It is essential to develop new standards in measuring and specifying these new properties.	
## 6.4 Non Conventional Technologies	
### 6.4.1 White-Bio Technologies and Bio refineries	S MT
Industrial or white bio-technology is the application for the processing and production of chemicals, materials and energy through enzymes and micro organisms to make products in sectors such as chemistry, food and feed, paper and pulp, textiles and energy. White bio technology could provide new chances to the chemical industry by allowing easy access to building blocks (e.g. succinic acid) and materials that before were only accessible via complex routes or not at all. White bio-technology may establish an effective way to use renewable resources, e.g. in bio refineries. An integrated and diversified bio refinery is an overall concept of a processing plant where biomass feedstock is extracted and converted into a spectrum of valuable products. Bio refineries combine and integrate necessary technologies from the biomass supply and conversion technologies through the core bio processing (white bio-technology) and downstream processing steps towards the final application of the use for society, therefore covering the whole industrial biotechnology value chain. Different technologies to convert biomass raw materials into industrial intermediates and	

Proactive Initiatives	**Priority**
consumer products should be developed. The main objective is to develop white bio-technology and bio refineries to reach a technological level on which these technologies will become as efficient and competitive as conventional technologies. This requires a lot of R&D activities such as production of chemicals and materials, which would otherwise not be accessible by conventional means, or production of existing products in a more efficient and sustainable way, increasing eco-efficient use of renewable resources as raw materials for the industry, biomass derived energy. These are based on biotechnology and can cover several areas such as: an increasing amount of our energy consumption, development/selection of raw material for specific applications, development of new procedures of enzymatic or chemical modification of biomass and transformation to monomers, development of new technologies for polymerisation, fibre pre-treatment and polymer processing, development of cost-effective, biodegradable biomaterials and bio composite materials and products made of them. The expected research activities should integrate a broad range of research competences.	
## 6.5 Management of Hazardous Substances	M MT
Value chain-based production activities have become more complex with new technologies, materials and interdependencies. Additionally, the requirements on production processes are permanently increasing with regard to efficiency and environmental friendliness. This collaborative project should have in focus economically and environmentally efficient cleaner production methods and control mechanism which control eliminate or reduce hazardous outputs. An 'integrated value chain approach', which involves avoiding, controlling, safe management and diligent use of hazardous substances including hazardous chemicals and hazardous wastes, is strongly encouraged. The aim of the project is to analyse the whole production process taking into account the occupational health and safety, the detailed risk assessment of hazardous substances in special process sequences and cleaner alternative methods and strategies to reduce hazardous substances. Emphasis should be on the safe management and development of ICT-based control mechanisms.	

# 7 Emerging Factories of the Future	
## 7.1 European Manufacturing Standards	M MT

Technical standards are common and have a wide range. The development of standards for efficient and sustainable manufacturing management should be a part of the strategic development of the European manufacturing. Several main standardisation activities and required standards should be mentioned:

- Standards of the European manufacturing management include
  - o Management of sustainability,
  - o Actualisation of quality and environmental management,
  - o Certification in global production and work: social areas, ergonomics, management systems, IP, security,
  - o Standards for the exchange of data and information in networked production.

The activity should be launched to define the requirements of standards, based on existing regulations.

7.2 Emergent Technologies for Manufacturing Systems

7.2.1 High Precision Manufacturing by Plug and Play, Components Based on Adaptive Smart Materials   M MT

The main objective of this topic is to create a new generation of active plug-and-play components, based on intelligent materials or combinations of passive and active materials (engineered materials) to increase the adaptiveness of production systems for changing conditions.

Proactive Initiatives   **Priority**

The intelligent plug-and-play systems can feature sensing and actuator structures, adaptive control and energy harvesting to allow a high accuracy of production systems under different conditions and to overcome traditional limitations on dynamics versus precision.

Research should focus on self-sufficient, self-sensing and self-actuating intelligent plug-and-play components based on smart materials. Such systems should easily implement and self-adapt their range of properties, depending on the changing process conditions. Technical key points are the compensation of static and/or thermally induced dislocations, vibration damping and the decoupling of oscillations. Vibrations could be used for energy harvesting processes to transform kinetic energy into electrical energy, to drive the intelligent system and keep European energy resources.

Deliverables include components and methods for intelligent, self-sufficient plug-and-play systems.

Radical new generations of adaptive production systems by means of active, self-optimising plug and play components based on multi-functional intelligent materials, can be achieved by implementing this innovative technology. Improved dynamics and a higher precision as well as a high level of reliability for changing process conditions represent outcomes of employing the technologies and tools in manufacturing engineering areas.

7.3 Knowledge-Driven Factories

7.3.1 Enabled Business Models for Manufacturing

S
MT

The reach and richness of the information and knowledge, available through global ICT services should significantly impact and support the transformation of European manufacturing industry. In particular, the service sector has adopted the use of this technology in interesting and competitive ways. Therefore, if European manufacturing is to embrace concepts such as the extended product service paradigm, in ways that would allow it having a pre-emptive position within global manufacturing, then it must adopt and understand new business models. These new models will almost certainly use ICT in novel and innovative ways.

However, with regard to ICT, manufacturing has unique and individual challenges. The knowledge can be more complex and requires much higher accuracy than traditional paper-based transactional data. For example, product and process data must be timely, completely accurate and have to take account of many historical configurations if they support the full life cycle. In addition, this complex data must be available as contextual data in easy-to-use formats to support business

Proactive Initiatives	**Priority**
processes as diverse as product design, process design, manufacturing, training, service and marketing functions. Finally, it should also capture the tacit knowledge, locked into products and processes in ways that allow effective, repeatable and systematic business models and processes to be developed and make it freely available. The research activities in this field aim at building on the strategic concepts of Europe at the centre of a global manufacturing capability, tacit knowledge capture and exploitation, lifecycle data management and new business paradigms supported by the reach and richness of state-of-the-art ICT. The conception and development activities should address a full transition/change cycle and include definition, design, development, test and prototyping phases.	
7.3.2 Implementation of Simulation Technologies Simulation is usually an analytic instrument to find out the behaviour of systems under the constraints of usage. It is used for planning and optimising the layout of logistic and manufacturing systems and the design of machines. Future capabilities of real-time control allow the integration of simulation in the systems to analyse the behaviour in relation to situations. This demands the integration of simulation systems in Manufacturing Execution Systems (MES) as well as in machine and process control. Feeded by sensorial supervision, monitoring and the actual load, it seems to be possible to look ahead on what happens and to compensate deviations of precision or to control manufacturing processes by learning from the future. The implementation of special methodologies like KNN, MD, Monte-Carlo and discrete models, seems to be of high interest for operating in instable parameter fields, inducing the increase of the efficiency of manufacturing systems.	S MT
7.3.3 Comprehensive and Holistic Approaches of Multi-scale Modelling and Simulation of Manufacturing Systems Modelling and simulation of complex networks of manufacturing systems is regarded as one of the central challenges for research in manufacturing technology. Adaptable simulation networks of co-operating and semi-autonomous software systems will lead to a fast market launch for new products as well as to shortened set-up times of production systems. The particular challenge of modelling ad simulation of manufacturing systems consists not only in determining the interactions of the different parameters and influencing factors but also in representing them in real-time simulation models. In the context of the different usages, the term 'real time' is understood here as a fast reaction to arising events as well as the time-deterministic calculation of plant behaviour for the control-coupled simulation. The	S MT

Proactive Initiatives	Priority
multi-scale simulation of manufacturing systems aims to connect different computation and data models based simulation approaches through a continuous information flow, higher order equation systems and universal interfaces. The goal of this connection is the bridging of the discrete and numerical simulation, and of the areas of application in the manufacturing domain, as well. 'Multi-scale' not only means the spatial and temporal scales within separate manufacturing processes, but also the different scales of all running processes in the whole manufacturing enterprise, called factory, as well as the different scales in the model itself. Therefore, for the purposes of the virtual representation of the factory as a whole, at different levels of abstraction, several heterogeneous modelling approaches have to be coupled.	
7.3.4 Holistic Factory Simulation In order to comprehensively approach the modelling and simulation at all scales of manufacturing systems, from network to manufacturing processes, the integration between the heterogeneous simulation models is necessary. This heterogeneity consists on the migration of simulation models from numeric simulation for the process and material modelling towards the discrete simulation for the purposes of logistics simulation. The complexity arises when approaching the modelling of the horizontal scales of the factory as a whole, beginning with the technical manufacturing processes, through equipments, robots, production systems, segments and network of production systems. As a conclusion, the following aspects and scales have to be regarded: ■ all levels of factory structures: manufacturing processes, machines, manufacturing system, and network of manufacturing; ■ several areas and concepts related to: mathematics, physics, chemistry, engineering, economic science; ■ several simulation methods: molecular dynamics, finite elements, event-oriented simulation; ■ spatial expansion: atomically, microscopically, mesoscopically, macroscopically; ■ temporal expansion: nanoseconds, seconds, hours, days, weeks, years. The application areas of multi-scale simulation for modern manufacturing systems lie particularly in: a) factory and logistics planning, b) work and process planning, c) construction of the operational funds, d) programming, control, e) processing and e) quality management. The concrete product emerges from the heterogeneous simulation models representing the prototype of the digital, virtual and real-time factory.	M/L LT

Proactive Initiatives   **Priority**

7.3.5 Manufacturing Process Modelling and Simulation All manufacturing processes are instable because of the high number of dynamic influencing factors and the technical problems to define all phenomena and their relation (interferences and overlaying) or to measure the real conditions and parameters in processes. High variances and the potentials of inventions are usually evaluated through physical experiments. Now high-performance computing and a high number of computerised methodologies, to realise process simulation in the areas of mechanical, coating, joining processes and non conventional processes, can be used. Processes are influenced by the thermal, mechanical and dynamic behaviour of machines and robots. And they are elements in process chains from the raw material up to the finished product. Simulation can even be used for analysing solutions, which are designed in CAD and CAM. Simulation reduces the cost and time of physical experiments and tests. This action focuses on the research of applicable modelling and simulation technologies in the fields of processes with mechanical, energetic, fluidic and chemical phenomena for modelling and simulation of parts manufacturing. The simulation systems should have links to CAD-models and integration of basic analytic methodologies for engineering finite elements, mechanics and fluid mechanics, molecular dynamics or others. They have to be integrated into the manufacturing engineering chains. The models have to be evaluated by experiments.	M MT

**7.3.6   New Classes of Models for the Simulation of Complex
        Manufacturing and Assembly Systems**
 	
The strategic planning of complex manufacturing and assembly systems is greatly in need of modelling tools that can be used to quickly and clearly interpret and evaluate ideas to reorganise and expand such systems.
    One mandatory attribute of the model class, which should be developed, is a capability to map change dynamics of order, product and resource quantities for a defined time horizon. Existing static models for mapping manufacturing and assembly systems fail to meet these requirements. The two well known classes of dynamic models – system dynamics models and microscopic simulation models – also fall short since system dynamics models are too abstract and microscopic simulation models too intricate for tasks of strategic planning.
    Input data of mesoscopic models include a planner's conception of the anticipated development of individual sub-processes, specified in quantity-time diagrams (QTD). Only total quantities of relevant objects (orders, products and resources) are mapped. Individual simulation objects are not. In this context, the term mesoscopic refers to the transition in modelling and simulation from movements of individual
	M
	MT

Proactive Initiatives   **Priority**

objects to spatially distributed movements of entire object groups that
are mapped, based on specific attributes. Even the primary results of
a simulation have to be furnished as a QTD, which can be used as the
basis to freely calculate definable parameters for evaluation. The
main research has to focus on the development of fundamentally new
classes of models, which support the aggregated representation of
processes in complex manufacturing and assembly systems. The core
of such models has to be mathematically formulated, because this
makes the necessary transformations of QTD possible. The model
classes, which are to be developed, will be the basis for developing
new possibilities in order to map and analyse processes, which planners will be able to employ to complete their strategic tasks for complex manufacturing and assembly systems with significantly more
speed and precision.

7.3.7 Modelling of Parallel, Serial and Hybrid Kinematics

S
MT

In recent years, machine tool builders are forced to develop machine
tools which enable the highly-effective and accurate manufacturing
of a wide range of different products with different manufacturing
attributes, e.g. parts which have to be milled and turned in one clamp-
ing. Beside these different technical specifications, machine tool
builders are forced to develop optimised machine tools, which are
quickly successful on markets with high cost pressures.

Under these circumstances it is extremely difficult for small and
medium-sized enterprises (SMEs) to choose the right machine concept, to define the integration of hybrid processes and to optimise the
kinematic structure and the structural behaviour in the early concept
stages. In addition to the complex decision of the machine's setup,
SMEs are often not in a position to use various complex software
tools to evaluate different concepts.

Within this project different easy usable methods and simulation
tools for the design process of serial, parallel and hybrid kinematics
of machine tools will be developed, taking into account the expected
range of products and manufacturing processes. The complex mecha-
tronical system machine tool should be represented by suitable sub
models, which enable companies to compare different concepts and
find the best possible setup of the machine tool under the mentioned
boundary conditions.

7.3.8 Virtual Reality-Based Simulations for Machine Operations and Life Cycle Impacts

S
MT

In the planning, development and utilisation of machineries and
plants, digital methods need to be consistently employed throughout
the entire life cycle to generate improvements, shorten planning times

Proactive Initiatives	Priority

and enhance product quality and reliability. It is becoming ever more urgent for enterprises to integrate advanced VR technologies in their existing infrastructures and to reorganise existing processes. Virtual functional models of specific machine configurations as the starting point, virtual products, machine and plant models are produced, which, additionally incorporate usability concepts and educational methods, already enable evaluating and improving products and processes in manufacturing and operating enterprises before their real commissioning. Functions and processes can be safely tested and procedures can be trained.

The research efforts have to initially focus on the development and integration of VR applications in the entire life cycle. These applications include planning, development and simulation tools, new methods and procedures for occupational health and safety; oversight and support of certification processes, development of process models and knowledge storage systems with supporting technical-didactic methods and training personnel in different processes on complex products, machinery and plants.

New methods and tools will be combined by using distributed simulation, methods for simulating and analysing shape varying and moving 3-D body surfaces, mixed reality technologies for the control, operation and maintenance of complex assets, augmented reality applications for process supports and quality assurances. The expected results are VR based tools, methods and prototype applications for machines, production systems and plants and production logistics networks.

7.4 Real Time Management

7.4.1 Built-to-Order New Models for Production Design, Planning and Control in Individualised Productions

M
MT

In the course of internationalised competition, production costs and time (e. g. delivery time, adherence to delivery dates) have become more important in relation to traditional quality targets. Especially small series productions with often highly individualised customer products are confronted with the challenge to cope with the conflicting aims of high efficiency, good process reliability and speed. Therefore, new models for production design, planning and control in individualised production are needed.

The main results in this field should provide mechanisms and solutions to improve small series production by overcoming the conflicts between the targets of efficiency, process reliability and speed. The research should aim at improving process designs and operations in a highly customer-individualised market environment, in order to

Proactive Initiatives	**Priority**

achieve good process reliability, short delivery times and low production costs at a time. The research should focus on small series productions in larger companies and SMEs which are typically struggling with poor economies of scale, high product variety and problematic material availability, as well.
 	
7.4.2   Build-to-Order in Manufacturing Networks
 	
European manufacturing enterprises have to compete based on a range of performance objectives such as quality, price, delivery, responsiveness and flexibility. Whilst manufacturers develop shorter product lifecycles and offer a greater variety of models, this provides shorter 'market windows' in which generating the sales volume is necessary to support the massive development costs of the new product. In most manufacturing sectors the current system is still mainly forecast, based on production and push-based selling, using discounts and incentives is leading to lower profits, thus more volume is needed to maintain the equilibrium. To face these challenges, companies and supply networks have to think about new production systems and manufacturing concepts which enable a fast response to the customer's needs and a flexible handling of capacities. Furthermore companies are moving or have moved from centralised operations to decentralised operations in order to take advantage of the available resources and in many cases, simply to be closer to their markets. Therefore, manufacturing supply chains undergo a major transition. Currently the supply systems are for the most part based on 'stock push', whereby the majority of products are sourced from existing finished goods inventory in the marketplace.
    Build-to-Order (BTO) strategies offer a new direction for manufacturers who suffer in this climate of spiraling costs and punctured profits. But these strategies have to be extended to encompass the changed global environment of the markets. Consequently, build-to-order strategies have to be investigated, designed and developed for network environments. This encompasses the ramp-up, production, and phase out phase of the product life cycle.
    European manufacturing industry faces fierce competition in all its major markets and is dealing with a customer who is more demanding. To face these challenges, companies and supply networks have to think about new production systems and manufacturing concepts which enable fast responses to customers' needs and a flexible handling of capacities. Consequently there is a need for BTO strategies for manufacturing networks.
    The main research targets are the creation of build-to-order manufacturing network processes and methods based on the application of enabling ICT infrastructures. Such network-wide BTO strategies require a close collaboration within business processes for
	M
	MT

Proactive Initiatives   **Priority**

capacity planning, order management, and production and transport especially if BTO parts delivered by first and second tier, suppliers are manufactured in BTO manner as well.

The concrete results of performing research and technological activities in this area represent new methods and business processes for network-wide BTO as well as the application of enabling ICT tools.

7.4.3 Supply Chain Integration and Real-Time Decision-Making in Non-hierarchical Manufacturing Networks

S/M
MT

The aim of the networked production is to cut logistics costs, to reduce high inventories of current assets and simultaneously to shorten the lead times of material and information as well as to improve the service levels in a customer-orientated way. These aims can only be reached by means of common efforts, which result in benefits for any company involved, when realised. In the already existing manufacturing world of highly distributed value-adding activities, which will further increase in the upcoming years, these targets can only be achieved by means of the close collaboration of the companies in the network.

Against this background it is obvious that a central planning and control of a network is only realisable if one dominant company is existent in the network, which is, however, rather uncommon. In fact, in most cases, the companies try to keep their acting and decision autonomy, which is essential for them for achieving their company targets. This inevitably results in a decentralised planning and control of manufacturing networks – although a central coordination of a network could probably be possible, regarding technology. Therefore, production in the future will predominantly occur in non-hierarchical company networks. The integration and production/operation management of such networks is characterised by a non-centralised decision making. Depending on the customer and the product, the rules and procedures for this decision making may change.

7.4.4 Optimisation with Knowledge and Cognitive Systems

M
LT

Capacity analysis and planning is a key activity in the provision of adequate customer service levels and the management of the company's operational performance. Traditional capacity analysis and planning systems have become inadequate in the face of several emerging manufacturing paradigms. One such paradigm is the production in collaborative enterprise networks, consisting of subsets of autonomous production units within supply chains working in a collaborative and coordinated way. In these distributed networks,

Proactive Initiatives	**Priority**
capacity analysis and planning becomes a complex task, especially because it is performed in a heterogeneous environment where the performance of individual manufacturing sites and of the network as a whole should be simultaneously considered. This collaboration can only work if the network companies working along the value chain are integrated through synchronised processes and harmonised IT systems. This vertical integration on the one hand means that the processes for the planning and control of production and logistics have to be closer interlinked between the companies, enabling the exchange of order, inventory, demand, and capacity information, and synchronising the processes initiated by this information exchange, resulting in innovative collaborative production and logistics processes. On the other hand the necessary ICT systems supporting the processes also have to be integrated so that the information exchange and the process synchronisation is enabled throughout the network. These integration needs call for tools making it possible to model, evaluate and realise the process integration and to shorten the time necessary for the IT systems integration.	
7.4.5 Self-organising Management Non-hierarchical company networks aim at enhancing the competitiveness of European manufacturing sectors by increasing the capacity of industrial SMEs to operate globally in an agile manner, in order to adapt to the rapid evolutions of existing and future markets. The supply chain integration and production/operation management of such networks is characterised by a non-centralised decision making. Depending on the customer and the product, the rules and procedures for this decision making may change. Furthermore, companies can be part of several production networks at the same time, thus making the planning, management and optimisation a very complex task. The main development issues and targets are collaborative planning, management and optimisation of production resources including production planning and capacity management in non-hierarchical company networks as well as distributed planning/scheduling models and supporting tools. Also methods and tools for material flow management across the overall network and the product life cycle, integrated production monitoring, offering order status information for the customer and the network, equipment monitoring and maintenance, integrated maintenance including real-time monitoring (design, implementation, operation) enabling new and protected services for the production equipment as well as planning and control of reverse logistics / recycling are	S MT

Proactive Initiatives   **Priority**

targeted. The new methods and the supporting tools must work in a decentralised manner enabling the participating enterprises to work in several production networks at the same time. The securing of information and knowledge should also be given a special emphasis, as it is a key to the success of such networks.

7.4.6   Networked Multimodal Collaboration in Manufacturing Environments

S
LT

For each kind of system, mainly ICT systems, a high degree of user-friendliness at the interface between a human and a machine is a decisive prerequisite for being generally accepted. The new manufacturing production systems and their applications are characterised by multi-modal interfaces as well as by the pro-active behaviour of the application system. Therefore, various interfaces must be sensibly combined with each other, and the interaction with humans must be perfectly adapted to the individual situation of the human. Specific challenges including, among others, the selection of suitable interfaces for specific applications, the dynamic changes of interfaces based on changes in the state of the human such as "experiences gained" or "accident", as well as the experience-based optimisation of such interfaces.

Multimodal data, including video (gestures, body motion, facial expressions, gaze), and audio (speech) constitute natural ways to build interfaces with machines. However, the combination of multiple data sources face a number of challenges, arising from their distinct nature (like asyncronicity) and their complex relations (modalities can at times be redundant, complementary or contradictory). The following main research directions are envisioned as being relevant for multimodal interactions. In the first research area or direction, the aim is to directly command a computer, machine or equipment via static and dynamic facial and body gestures (a described hand trajectory, or a sequence of hand postures) and speech commands. The machines should become able to register human emotions (and related states), to convey emotions (and related states), to "understand" the emotional relevance of events. The second research follows the main idea to automatically analyse natural individual actions or group interactions and react accordingly in "smart" spaces like meeting rooms, or mainly in the so-called "Smart factory". The main research objective in the area of multi-modal interaction represents the development of an integrated framework for networked multimodal collaboration in manufacturing environments. The environment aims at sensing the existing computing and manufacturing environment and adapts them to provide a prescribed quality-of–service, by involving the information transformation, where it is needed, and evolving methods for multimodal

Proactive Initiatives   **Priority**

human/machine communications, implemented at client stations, in order to enhance naturalness, ease-of-use and functionality.

The implementation of the new human-machine interaction has to approach the following directions and aspects:

- Development of new statistical approaches for gesture recognition and enhancement of already existing and suitable approaches, namely Artificial Neural Networks (ANNs) and Hidden Markov Models (HMMs). The existing models, for ex. HMMs, well known methods to model temporal data do not exhibit optimal properties to discriminate between sequences of distinct classes.
- New approaches for the problem of data fusion for multimodal person localisation, tracking, and action recognition. The already existing and current approaches focus on the Graphical Models and the Sequential Monte Carlo for audio-visual speaker-based tracking and action recognition. There is a need to develop new asynchronous HMMs, capable of manipulating and managing several streams of information that could contain asynchronous joint information.

7.4.7   Real-Time Network Visibility by Mobile Components in Production Networks

S
LT

A necessary prerequisite for optimising and managing production networks continuously is the realisation of a real-time network visibility. With this visibility companies within the network have instantaneous access to the current network status in order to see where activities are not carried out according to plan, thus endangering production processes requiring the in-time materials supply and consequently breach delivery promises given to customers. This can be achieved through the application of radio frequency identification (RFID) technology. RFID tags allow the storage of important information like product identification, order information, delivery targets on a chip which can be read contact free. Furthermore the tags allow also that further information is written on them, thus making a real-time picture of the network status close to the product and moving through the network is possible. By attaching RFID tags to the parts, products or transportation containers, reading and writing information on the tags and a full-scale visibility of the network can be realised, together with the possibility to excel the control of the network. This traceability can only be realised if all network participants, manufacturing and service companies, are integrated into an overall process approach, delivering the right

Proactive Initiatives	Priority

information on time to the other network companies. In addition to established technologies for the localisation of products like GPS or GSM between participants in the network, enable the development of radical new approaches for the monitoring and failure management in production networks.

A further research target can be found in the necessary integration of RFID systems into the different backend IT systems as enterprise resource planning, production, transportation and warehouse management. Only with this integration the network visibility together with its advantages for industry can be achieved.

Finally a scalable filtering, compression and visualisation of the captured network status, adapted to the different needs and levels of the network companies, has to be realised in order to make the utilisation of the gained network status information as efficient as possible.

Networked production requires the development of new processes for network-wide monitoring and exception management through the application of new identification, communication and positioning technologies which can be mounted onto the products and parts of them routing through the production and logistics network. The manufacturing networks will employ location services to meet specific business needs such as preventing loss or thievery of valuable mobile assets, automating workflows, managing inventory and tracking equipments, supplies or people. Another important research target is to develop methods and supporting tools for the filtering of the available information so that human problem solvers are immediately guided towards those objects – parts, production equipment, and transportation devices – that require their immediate attention. Research will focus on the application of smart mobile components and networks integrating multiple wireless communication technologies (GSM, GPRS, WLAN, RFID, Bluetooth) and sensors as well as integrating them into intelligent manufacturing structures. Deliverables shall include processes, methods and the application of supporting tools for a mobile business system suitable for production networks. A demonstration and validation in production and logistics network is required.

7.5 ICT System for Knowledge-Based Products and Processes

M
LT

Manufacturing systems are mechatronic systems. The engineering of customised manufacturing systems includes the engineering of mechanics, electrics and electronics. The reliability of systems depends on the complexity of integration. Engineering needs the integration of systems for the design and analysis of integrated data

Proactive Initiatives   **Priority**

models and the systematised engineering of software. The integrated engineering of complex technical solutions is a strong recommendation in industrial companies for:

- Modular design
- Reproducibility and adaptability of software components
- Standardisation of components
- Efficient generation: function oriented
- Test equipment and simulation

The main research in this area has to be directed towards the development of engineering systems for customer specific software and for manufacturing (machines, tools, transport, handling, flexible automation), aiming at managing the complexity and increasing the adaptability and reliability.

7.5.1   Capturing and Synchronising Heterogeneous Production Data with the Real Time Digital Factory

S
LT

The real-time factory offers an intelligent, real-time operational management of factory processes and resources. It tightly integrates the real factory with the virtual factory by continuously communicating, connecting and evaluating the factory's operational data. The real-time factory additionally introduces interconnected, self-adapting, cognitive devices and systems for the real-time operational management of the factory. This new way of factory planning employs a set of various local and distributed planning activities (e.g. production facilities, logistics, organisational planning) for a digital factory and makes predictions for a virtual factory which are based on the realisation of the original plans. The technique of collaborative and distributed planning addresses the generation of plans and of monitoring instructions at different spatial and temporal events.

The real-time factory is based on sensors which collaborate in complex networks, continuously acquiring the actual state of the factory, e.g. machine states, flow of material, product quality, and human resources. By using ubiquitous computing techniques and self-organising sensor networks, data is collected, aggregated and processed in an intelligent way. These data are integrated and managed in a repository forming the basis of context-aware systems in the real-time factory.

Uncertainties occurring in dynamic and distributed environments have to be handled with stochastic methods (e.g. Bayes-nets, Markov Decision Processes - MDP, Partial Observable - MDP) supported by collaborating multi-agent technology. This supports the operation in highly dynamic environments since production

Proactive Initiatives	Priority

plans can be repaired online or even be newly generated by context-aware sensor data interpretation and communication protocols. Due to the economic relevance those processes must be accompanied by adequate valuation, controlling and planning models. The dominant operating entities of the Stuttgart Enterprise Model are people, therefore the paradigm of distributed cognition will not be applied in order to separate man and machine but to consider both as distinguished units in a socio-technical system as a whole.

7.5.2 Distributed Tele-Presence in Haptic-Visual-Auditory Collaborative Manufacturing Environments

S
LT

The rapid advances in computer and network technologies induced the challenge of distributed and collaborative design environments. This is based on the more increasing importance of the remote collaboration for manipulating 3D models in the design domain, due to the competitive and complex product development processes. HCI community envisioned that tele-presence represents on of the main enabling technologies, having a relevant impact on the remote collaboration in manufacturing engineering. Tele-presence means technical tools enabling a human operator to be present in another, removed or not accessible remote environment with his subjective feeling. Supporting verbal and non-verbal communication is seen as an important issue for facilitating the remote collaboration and a main requirement for the implementation of distributed 3D collaborative design environments.

The research in the tele-presence field aims at overcoming several challenging barriers between the operator and the tele-operator, for ex. a barrier can be the distance, but also the scaling (small-scale tele-presence - e.g. minimum invasive surgery, micro assembly - or also large-scale tele presence). Additional to the visual and acoustic sensory immersion, in particular haptic immersions are required, mainly tactile (pressure, temperature, roughness, vibrations...) and kinesthetic (proprioception, inertia effects, the force of gravity) channels are used, in order to improve the immersion in the virtuality.

The overall objective of this area represents the design and implementation of a so-called haptic-visual-auditory collaborative manufacturing environment, seen as collaborative common work space, in which distributed tele-presence collaborations, with the modalities mentioned, are supported (Tele-collaboration). Networked and geographically distributed operators in a common manufacturing environment have to solve a complex manufacturing task and they are supported in their perception and immersion in the virtual environment by auditory, visual and haptic immersions.

Proactive Initiatives   **Priority**

The research program can be oriented towards 2 areas:

1. Development of the knowledge base, which is required for the design of the haptic-visual-auditory collaborative manufacturing environment, consisting of new models, methodologies and enabling technologies and applications. It concerns methodical bases for communication, for the different visualisation and immersion modalities

2. Development of different employment scenarios where the haptic-visual-auditory collaborative manufacturing environment can be used.

The integration of envisioned haptic-visual-auditory systems and tools improves the overall communication and collaboration between the involved operators in the manufacturing processes.

7.6   Technical Intelligence by Machine Learning in Manufacturing (Learning Factory)

M
LT

The application areas of pattern recognitions like image analysis, character recognition, speech analysis, man and machine diagnostics, person identification and industrial inspection are converging in the manufacturing industries in the last years. Studying the operation and design of systems that recognise patterns in data, and inclosing subtopics like discriminant analysis, feature extraction, error estimation, cluster analysis (together sometimes called statistical pattern recognition), grammatical inference and parsing (sometimes called syntactical pattern recognition), the pattern recognition area supports more and more manufacturing enterprises, mainly for cost reduction purposes. The cost of manufacturing in general, and in particular the costs associated with assembling complex devices, would benefit tremendously from this kind of technology.

The overall objective is the development of new and innovative pattern recognition methodologies and tools for the purposes of manufacturing engineering. The research efforts can be directed to several areas and validated in several specific application fields:

•   Development of pattern recognition methods and tools for the 3D design activities. An envisioned application is to suddenly recognise the orientation of a part in a bin. By using the pattern recognition, the operation becomes easy, and orders of magnitude cheaper than any AI solution, if the corners of the part announce themselves (requiring a couple of localisers on each part).

Development of a pattern recognition-based integrated approach to design compact manufacturing facilities by using facility layout

Proactive Initiatives	Priority
planning techniques to drive design and selection of multi-function machining centres. This is supported by the integration of pattern recognition systems and tools into the collaborative manufacturing environment by having as a proposed application field the integration of facility layout and flexible automation, as two approaches for reduction of material handling costs and product travel distances. These two have been always implemented independently of each other.	

7.7 Computing Systems and Embedded Platforms for Advanced Manufacturing Engineering

### 7.7.1 Pervasive and Ubiquitous Computing for Disruptive Manufacturing	S/M LT

A new challenge in manufacturing engineering represents the migration of ubiquitous computing into the manufacturing world. The manufacturing engineering community and enterprises need this emerging new technology, ubiquitous computing, in order to develop an adaptive and evolvable manufacturing environment which is present anytime, anywhere, and which can access desired information by real-time. This will support the globalisation of business environments which globalise the production and logistics issues across multiple factories over geometrically remote sites, as well. This rapidly changing global business environment requires each manufacturing enterprises to remotely monitor and control from real time status of processes, materials, production procedures and workers for an optimum management.

This ubiquitous environment for manufacturing engineering needs to economically provide diverse communication functionalities to interface with existing (e.g. Process I/Os, serial communications and fieldbus) and new (e.g., industrial wireless Ethernet, RFID, USN (Bluetooth & Wi-Fi) and PLC (Power Line Communication), data processing and decision making functions. It is also essential that the new environment needs to use tether-free internet technology. This technology will enable information integration of all applications being used in manufacturing systems and the best use of the information for optimum decision makings.

The research activities aim at supporting the manufacturing enterprises with solutions to collect and use production information from globally distributed factory by real time.

The scientific goals and planned research steps are mainly represented by the development of a ubiquitous environment for advanced manufacturing engineering that integrates processes, manufacturing

Proactive Initiatives   **Priority**

resources as materials, production procedures and equipments, machines and people (workers) within a single machine-to-machine
(M2M) physical platform under a centralised management to establish the ubiquitous manufacturing;

7.7.2 Grid Manufacturing: Advancement of Grid Computing for Manufacturing Purposes

S
LT

Networked manufacturing systems are expected to have much more
flexibility to respond to dramatic changes in the world market. But a
real responsiveness might come from dynamic and unlimited resource accessibility rather than from a rigid company boundary.
Grid manufacturing is regarded as the next generation advanced
solution for the bottleneck of networked manufacturing. With the
appearance of grid computing, whose core is dynamic and has a
cooperative resource share, grid manufacturing will be the further
goal of networked manufacturing. Grid manufacturing realises the
share and integration of manufacturing system resources and supports the collaborative operation and management. It is based on
grid and associated techniques to supply enterprises with reliable,
standard, easy accessible and cheap manufacturing resources and
services, thus realising the cooperation of the whole process including supply chain, design, manufacturing and sale, and decrease the
costs, shorten the manufacturing period, raise the quantity and finally improve core competences. Current challenges for the network
of manufacturing enterprises: innovation, speed and flexibility.
Required characteristics of networks of manufacturing enterprises
are: collaboration, decentralisation and inter-organisational integration. In order to turn this new concept into reality, few definitions
have to be mentioned, as follows:

A) grid manufacturing = a new industrial paradigm, aiming at
   enabling the collaborative sharing of resources and competences in manufacturing engineering
B) grid manufacturing = the use of networked and distributed
   manufacturing resources in order to:
   - support the collaborative planning, operation and man-
     agement of manufacturing,
   - respond to the emerging challenges: innovation, speed
     and flexibility"
C) grid manufacturing = a common collaboration platform for
   networked manufacturing
D) grid manufacturing = integration of grid technology with
   manufacturing technology for the purposes of networked
   manufacturing.

Proactive Initiatives   **Priority**

The overall objective of the research performed in this area represents the development of a concept for grid manufacturing paradigms and of the corresponding roadmap, having as an expected overall output the first prototypical implementation of a Virtual Grid Manufacturing. The envisioned scientific goals and research steps are as follows: • Identifying the needs and the development of new business models and rules, required for inter-enterprise collaboration, rules of trust, • Identifying the needs for the development of standards to enable the manufacturing enterprise to exchange its products, services, to develop the interfaces with others, related to IP, ICT, financial, material flow level and other aspects, • Development of methods and tools, required to assist the manufacturing enterprise: to be connected to the grid, to provide products, manufacturing operations and services to the grid, to operate in the grid, mainly to access the operations and services offered by the grid, • Design, development and prototyping, the first broker of grid manufacturing operations and services under the name Manu-Google, as a similitude with google. As main outputs of the above mentioned activities, following concrete expected deliverables have to be mentioned: • Package of recommendations concerning the required standards, enabling/facilitating the participation of manufacturing enterprises in the grid, • Package of recommendations concerning the required tools, enabling the grid operations, • Results of studies in several main aspects of the grid, • Potential architecture of the grid, • Practical simulations and demonstrations, • Roadmap for grid manufacturing. **7.8 Disruptive Factories** **7.8.1 "Bio-nano" Convergence** Many consider that the convergence of the bio and nanoworlds will be a rich source of new products particularly for human health. Products emerging from the science base are likely to form the basis of new industries. Such multidisciplinary industries require effective new product introduction processes and tools, and new manufacturing	 S LT

Proactive Initiatives	**Priority**
processes and production systems that are both effective and match global regulatory requirements. Many will require new businesses and models and delivery methods. The main development issues and targets are: • Tools for the commercialisation of products emerging from the science base at the convergence of bio-nano. • Business models, new product introduction processes and technologies for the delivery of bio-nano products. • Processing of current and emerging naturally derived and synthetic medical device, therapeutic and industrial biomaterials. • Step change methods/disruptive processing of chemical pharmaceuticals of increasing complexity. • Scalable processing of bio-pharmaceutical and genetic, cell, tissue and regenerative and nano-medicine-based therapies including third generation tissue scaffolds. • Sensor, instrumentation, measurement, characterisation and control techniques and systems for the above mentioned, including bio-chips and laboratory on a chip technology. The expected outputs are: new generations of products and manufacturing processes, new business models and methods for delivering these products, and instrumentation and characterisation systems for these emerging products. **7.8.2 "Bio-cogno-ICT" Convergence** The modern scalable, adaptable, responsive manufacturing enterprise, the so called factory, has to be supported along its life cycle phases by the newest convergent technologies, mainly by bio, cogno and ICT. So, it is "cognitive" at all its scales (network, manufacturing system...), by embedding elements of technical, social and distributed cognition. It has to be "consciously clean" by an employment in critical phases of the environmental technologies. The enabling ICT technologies, like autonomous computing, ambient intelligence, or web-services, are still far from meeting this challenge as the only ones of the manufacturing industries. A new engineering approach is required, having as a main foundation the conventional and new manufacturing technologies, and as pillars, the nano, bio, cogno and information and communication technologies, which converge enabling each other in the pursuit of this common goal: to make the envisioned "next generation, conscientious clean disruptive factory" real. This new engineering approach bases on concepts and methods from the interdisciplinary field of cognitive science, mainly represented by artificial intelligence, mechanical	 M LT

Proactive Initiatives   **Priority**

and electrical engineering, biology, cybernetics, psychology, linguistic, neuroscience, social sciences and philosophy. The employment of convergent technologies disrupts the traditional way of approaching the factory, in its economic sense, by enhancing it with the "disruptive" feature.

The main objective is to harmonise cogno, bio and ICT, under the orchestration of manufacturing technologies for developing innovative concepts, models and various implementations of the main issues of technical, social and distributed cognition in different sociotechnical environments, mainly focusing on the manufacturing systems or factories. New concepts and paradigms for "cognitive technical spaces", adaptability, safety engineering, usability, scalability, robustness and technology acceptance etc. are proposed to support sustainable development of the European manufacturing sector. The planned research activities lead to overall design processes and generic models that are used in all application areas.

The research activities are conducted to develop concepts, models and methodologies/tools for design and manufacturing networks of cognitive manufacturing machines, such as prototypical implementations of robust and adaptive cognitive manufacturing systems for the following application areas: design of cognitive assistant systems, products such as cognitive cars and cognitive traffic control, cognitive robots, machine tools and production control, cognitive systems for domestic and organisational environments.

The main research areas which serve as a fundamental basis for achieving these concrete goals, the design and construction of several instantiations of cognitive technical systems are:

- Technical, social and distribution cognition,
- Modelling, simulation and prototyping of cognitive systems,
- Human and machine learning in cognitive systems,
- Communicating, perceiving and acting in networks of technical systems,
- Cognitive systems, safety, reliability, security and comfort engineering.

Deliverables will take the form of prototypical implementations of the "Disruptive Factory" in industrial settings, in order to prove the migration of the new paradigm in the real manufacturing industry. The pilot prototypes would represent a valuable incentive for the private sector of investment.

8 Global Leadership

8.1 Science-Based Entrepreneurship Is Leading to Global Manufacturing

S
ST

The science-based models, methodologies and tools provide a rich source of potential products for manufacturing industries. Some products emerging from the science base may even lead to techno-logical innovations of such a scale that they become significant industries in themselves. The liquid crystal display is a historical instance of the latter. The creation of radical new products takes longer than incremental conventional products and it is a higher risk. This requires corresponding radical approaches to investment and the return on investment. Innovators in science frequently lack the entre-preneurial skills necessary to focus on getting the best ideas to mar-ket and have little understanding of the requirements for enabling process technologies to allow their products to be realised at an ac-ceptable cost.

In order to exploit the new ideas arising from the science base by European businesses, it is critical to understand how such radical product developments can be financed and how values are developed and returned to investors during the early stages of the life of such new and exciting products. Radical physical products will demand new materials and new manufacturing technologies. The new product development process and its leadership demand a new type of entre-preneur, which is capable of interfacing between the enthusiasm of the science base and the pragmatism of the intellectual property pro-tection, finance and manufacturing. New business models for the creation of new generic materials and manufacturing technologies for emerging products are also required. Mechanisms of accelerating the pace of commercialisation of products from the science base and of increasing the knowledge of the science base on the emerging prod-uct requirements of the European economy and society must be found.

Proactive Initiatives   **Priority**

## 8.2 Competition in Global Manufacturing In today's global marketplace, companies face intense competition and increasingly sophisticated consumer demands. Innovation lies mainly within manufacturing networks, constituted by OEMs and SMEs, which compete on the market. The supply value chains are shifting from internal process management and cost reduction, typical of Lean Production, to value external collaboration, flexibility, and risk management (including continuity of supply), in order to achieve multiple sources of higher value. New business models are fundamentally required for compete in global manufacturing, making SMEs factors contributing to the European manufacturing success, reconsidering the nature of OEMs – SMEs, suppliers and partners. The technical content of these special investigations is focused on competition in global manufacturing to identify and exploit new opportunities for maximising values in manufacturing networks for European strategic sectors, such as Automotive and Aerospace. The challenges are to evolve the supply chain concept into a new net-centric approach; deploy a network approach to manage business processes and technologies; to develop a virtual "network" extension of organisations' internal capabilities; to build new levels of visibility and interoperability into organisation extended operations. These investigations aim at identifying the actual situation of key OEM-SME networks, regarding the product life cycle in the enlarged Europe; in order to identify the major European barriers to achieve a competitive network-centric manufacturing environment; to recognise the different role of partners in different OEM-SME networks; encourage diverse public and private stakeholders as high value partners in the OEM-SMEs network manufacturing to increase capability. These special investigations in global manufacturing competition aim at: analysing the Europeans best practices in the manufacturing network-centric for consumer goods, semi-finished and capital goods; benchmark by U.S.A. NACFAM) to understand 'how' organisations collaborate and not 'how much'; identifying the OEM internal and external business transformations, including "reactive" and 'proactive' integration in product development. Possible results have to: deepen the knowledge of the weak parts of the OEM-SMEs in order to provide an impact on SME networks: outline new business strategies, new tools and techniques to comply with product inside the OEM-SME networks; empower the European manufacturers, within OEM-SME networks, in order to avoid, reduce and fill the gap with the U.S.A., Japanese and world counterparts.	S ST

Proactive Initiatives   **Priority**

8.3 Global Networking

8.3.1 New Business Models for Networked Virtual Factories

S
ST

Demands from the market, changes much faster in the future, so companies can only be compatible in these markets, if they develop, produce and distribute new products much faster than today. For all companies this is a problem of resources, especially for SME´s. To combine the strength of different manufacturing enterprises in the whole process and to be much faster – from the idea to the product – represents the main idea of the Virtual Factory. To organise such virtual factories, there is a need to approach all classical functions of enterprises on a higher level – on a virtual level and for networked virtual factories. As a consequence, new organisation models have to be developed, aiming at developing such self-organised and market-driven entities, consisting of models for:

- Market environment for new businesses (electronics platforms / stock exchange systems)
- Partner generating modules
- Optimisation – who is the best in which stage of the process
- Rules and rights
- Information and knowledge processing
- Product processing.

The development of a market driven organisation structure with specific aspects of indicators for combining competencies, rules and rights, information systems and logistic aspects has to conduct towards the development of a new market model for networked virtual factories.

8.3.2 Management of Global Networking

M
ST

Globalisation of work is one of the leading trends in the business world. To transfer the ideas and rules of the European management system helps European companies to cooperate with companies of other parts in the world. The European manufacturing standards can be transferred around the world, so that Europeans can also transfer the use of these standards in management methodologies, production processes and information systems for cooperation and working together to other companies. Global networks are useable for supply chains, networked companies and virtual factories.

8.3.3 Simultaneous Engineering in Open (Global) Networks

S
ST

More and more decreasing innovation cycles have been established in almost all types of businesses and markets. The emerging pressure on

Proactive Initiatives   **Priority**

products as well as production and logistic planning, regarding time
and quality, can be encountered by a solid parallelisation and reorgani-
sation of the engineering processes, the so-called "Simultaneous Engineering". As a consequence, the involved parties are facing immense
challenges, particularly in the field of design and planning processes
and the appropriate organisation as well as at the use of modern technical planning methods. These needs are intensified by market and
company globalisation, together with major enterprises' strategic decisions under the heading of "outsourcing". The respective engineering
processes are distributed over departments of a single enterprise as
well as across enterprise borders. Thus, the engineering in networks
has to be performed in a collaborative fashion. In this regard concepts
such as the Digital Mock-Up (DMU) for the product development as
well as the Virtual Factory and the Digital Logistics for the production
and logistic planning gain importance. The main requirement of such a
collaborative planning is the ensuring of the interoperability over the
complete engineering process.

The trend towards an international division of labour within production networks will significantly change the engineering of products and
the design of the respective production processes. Therefore, the respective engineering departments of the network's enterprise have to collaborate closer and faster in order to meet the increase requirements,
especially in respect to time-to-market and product customisation.

This collaboration of different engineering disciplines requires interoperable methods supported by appropriate tools. Thereon clear
organisational structures and processes of the collaborative engineering must be developed by defining activities, responsibilities as well as
rights and duties of the ones involved.

8.3.4 Innovative and Efficient Networking of Supplier and Customer Systems and Processes

S
MT

Concentration on the core competences and on outsourcing is parts of
strategic orientations in series production. In this area, the logistic
chains are industrialised and efficient. Companies with small series
and high diversity of variants have high problems in the management
of the chains: low volume, last minute changes, complexity of products, special technologies. In consequence to this, a high potential of
synergy is not activated and causes uncertainness in the process chains
and quality. Deviations from scheduled plans bring high losses in the
efficiency and markets. Solutions and methods are required to activate
the potential of synergy in the chains of cooperation: engineering,
supply of material, supply in manufacturing equipment, and supply in
services.

Proactive Initiatives	**Priority**
Possible solutions have to be developed and evaluated under real existing conditions: • Robustness of the cooperation chains • Reduction of organisational inefficiencies • Open architectures for information supply • New methodologies for order management (situation based) • Worksharing • Networking and neighbourhood • Regional networking in competences and capacity balancing • Implementation of innovative technologies • Standards • Human relations The focus of this action is the development of new solutions for efficient networking and the implementation in different sectors like machine-industries, manufacturing of capital intensive products.	

8.3.5 Global and Real-Time Network Management
 	
The overall vision of the global real time network management can be seen in the ability to have a real-time visibility onto all of the network's segments, acting locally on disturbances or changes in the demands, integrating the planning processes of all network companies efficiently and flexible, and enabling the adaptability of the network from the operational level up to a structural level of product and network design. This includes the definition of new business models, contractual arrangements, collaboration incentives and integrated sales networks. Business models will consider networks consisting of SMEs and large companies as well. Successful collaborative models coming from Europe, but national and regional experiences will also be considered.
    For this purpose new solutions will fully integrate different technologies (e.g., sensors, radio frequency identification devices, localisation devices, remote monitoring and control equipments) for planning and controlling material and information flows across critical processes in networks. The deployment of these solutions will consider the users' needs and requirements related to data and information accessibility and security.
    The need to co-operate with globally dispersed partners has made the management of production networks a very demanding task. Global manufacturing networks call for the ability to have a real-time visibility onto all of the network segments in order to act locally on disturbances or changes in the demands, and integrate the planning processes of all network companies efficiently. A global real-time network management requires further integration of sensors into the
	M
	LT

Proactive Initiatives	Priority
production and logistic equipment, collecting data about environment conditions and storing this information for the further decision processes on the local level. The design and application of advanced equipment, capable of detecting other devices, being able to communicate and thus enabling self-organising sensor networks for the solution of local operational decisions, will increase the scope of adaptability of the global production and logistics networks. Based on the achieved transparency of the network status, a continuous monitoring of the network performance is possible, opening the way to detect unplanned, delayed or missing events within the network very early and analyse whether the effects of these events are critical for the network operation. The envisioned outputs represent integrated solutions for the management of global production and logistics networks. These solutions must be demonstrated and evaluated in industrial settings. **8.3.6 Global Platform of Networked Service Management** In the future, traditional hierarchical and tight supply-chains will have to be much more re-configurable, agile, collaborative and responsive, moving towards a self-forming supply-chain and inevitably posing new and demanding challenges on its management. To support this envisaged trend it is necessary to proceed with the development of a conceptual framework for a self-forming business networking environment based on the idea of an innovative Plug-and-Do-Business Paradigm. Thus, it will be very important to explore different directions, namely: Support for short-lived ad-hoc virtual formations of collaborating partners and the issue of an enterprise to discover potential business partners upon demand and advertise itself in standard ways and the support of a highly dynamic involvement of an enterprise in different business activities, serving different roles at the same time. For producers of machinery and equipment, this means that the selling argument of the future is not any longer just the technical level of the product but rather its contribution for solving a problem of the user. This ability to solve problems of the user manifests itself in additional "added-value services" that assist the user with planning and dimensioning, rapid installation, smooth operation and uncomplicated system alterations. In short: value-added services covering the entire life cycle of the product. In order to achieve the flexibility and the resulting adaptability of a network, each company in the network has to define and offer services for the design, planning and control of the network segment, on which it is responsible for, based on their capabilities and competencies. Bringing these services into a service-oriented architecture for manufacturing	 M LT

Proactive Initiatives	Priority
and logistics planning and control applied wide networks, generating the necessary flexibility towards adapting to fast changing market demands by restructuring the product or the network, changing the network operations or using potentials existing in the current network status. Furthermore the realisation of such a service-oriented architecture supports the further decentralisation of activities in non-hierarchical networks, as the encapsulated services can be defined and executed by each company independently. Products are slowly loosing their dominate role concerning the market's success of production companies. Instead the market is demanding for 'all-inclusive' solutions incorporating the product itself as well as product-related services such as transport, installation, training, diagnostics, maintenance, and recycling. Only very large companies can offer all these services on a global level. The vast majority of manufacturers have to collaborate with local enterprises when offering their services to new markets. The establishment and the operation of such new collaborations require new methods and supporting tools for service offering, service discovery and service management. Research projects should cover and support the entire range of processes within networks for creating value added services and the interoperability of services.	
8.4 Innovative Customer-Driven Product/Service Design in Global Environments Intelligent customer driven innovation focuses on the integration of customer influence in the design and development process and the related demands of the manufacturing and logistic processes. Moreover, multi-site and multi-nation product development is becoming more and more an international business. Companies will design products, including production systems and even factories themselves, for customers all over the world. They will also develop and manufacture these products with partners and suppliers from all over the continent. Intelligent customer-driven innovative product service design in a global environment is set for new challenges, such as culture, specific customer preferences, location, production technology and logistics, round the clock 24-hour collaborative development, different cultures, attitudes and procedures for participating companies. The expected results can take the form of: validated tools for cost-effective and rapid creation, management and use of complex knowledge-based product services combining the customer-driven approach with enablers for competitiveness at internationally networked locations; tools facilitating collaborative design in temporary partnerships; and new business and management processes in virtual company networks around the world.	S MT

Proactive Initiatives   **Priority**

8.5 Global Security	
8.5.1 IP Security in Networked Manufacturing The activation of synergy potential in networked manufacturing is one of the strategic aspects of manufacturing industries. The exchange of knowledge is one of the critical factors in the collaboration and needs the protection of knowledge. The specific actions of IP-Security have to be elaborated in a coordinated action.	S ST

8.5.2   Innovative Methodologies for Protecting Intellectual
        Property and Know-How
 	
In recent years, innovative enterprises are confronted with technology know-how thievery and product imitations at alarmingly increasing rates. Especially SMEs are facing problems in enforcing legal and legitimate claims on their intellectual property and know-how. Since intellectual property rights seem to become less and less effective, a more holistic approach on know-how-based competitive advantages in technology-driven firms is needed.
    The research performed in this area aims at reducing the risk of European companies to suffer from illegal and illegitimate use of their intellectual property. This will be achieved by the development of advanced and new protection mechanisms, by an integrated assessment of their benefit and effort relations or through supporting companies in the selection of adequate solutions. Projects should focus on enabling new technologies, not on legal actions against imitations. Projects should address the current situation especially of small and medium-sized enterprises (SMEs) which is mainly characterised by low resources for the identification of imitations, legal action against them as well as persecution and lobbying.
	M
	ST

9 Education and Training in "Learning Factories"	S/M ST

9 Education and Training in "Learning Factories"

Technical and organisational innovations change the structure of manufacturing industries. Main drivers are new technologies for micro and nano-scaled products, engineered materials and new processes characterised by fast adaptation, networking and digital factories. The content of this action is the fast transfer of basic knowledge from research to application by education in learning factories. The learning factories have to be equipped with an integrated system for manufacturing engineering with 3D CAD, analysis and planning tools for manufacturing processes, with high end product data management, VR-systems and a physical laboratory with changeable manufacturing and assembly systems. The labs should even be equipped with new solutions for information supply like ubiquitous computing, wireless technology and navigation systems, implemented in an ERP, order management and manufacturing execution system. Simulation of logistics, kinematics and processes are elements of the learning factory. For education and training it is necessary to link the shop level systems with the digital environment. The Learning Factories offer basics for engineers and technicians in praxis with the following topics:

- Basic knowledge in changeable production systems,
- Optimisation of manufacturing in real and digital environments,
- Learning fast adaptation of factories,
- Usage of high end ICT in manufacturing,
- Management of change, from conventional to high performance technologies,
- Process planning and process management.

The learning factories are regional-oriented with relation to the structure and technology portfolio of the dominant sectors. The courses qualify for advanced engineering and management and participants should get a certification for the results.

ANNEX III

List of Contributing Organisations

AGORIA, The Federation for the Technology Industry, Belgium
Chris Decubber

Centre Technique des Industries Mécanique, Directorate for Regional and International Development, France (CETIM)
Daniel Richet

Daimler AG, Germany
Heinrich Flegel, Eberhard Bessey

European Committee for Co-operation of the Machine Tool Industries, Belgium (CECIMO)
René Groothedde, Oliver Cha

FATRONIK, Spain
Rikardo Bueno

Festo AG & Co. KG, Germany
Peter Post, Christoph Hanisch

Fraunhofer Production Alliance (FhG-VP), Germany
Engelbert Westkämper

Fraunhofer-Institut für Fabrikbetrieb und -automatisierung, Germany
(Fhg IFF)
Head of the Institute: Michael Schenk
Stefanie Germer, Jacqueline Görke, Daniel Reh, Barbara Schenk, Ulrich Schmucker, Antje Tiefenbach

Fraunhofer-Institut für Materialfluss und Logistik, Germany (FhG IML)
Head of the Institute: Axel Kuhn
Bernd Hellingrath, Sven Krause, Georg Pater, Markus Witthaut

Fraunhofer-Institut für Produktionsanlagen und Konstruktionstechnik,
Germany (FhG IPK)
Head of the Institute: Eckart Uhlmann
Philip Elsner, Pavel Gocev, Markus Rabe

Fraunhofer-Institut für Produktionstechnik und Automatisierung, IPA,
Germany (FhG IPA)
Head of the Institute: Engelbert Westkämper
Holger Barthel, Carmen Constantinescu, Michael Eisele, Harald Holezek, Michael
Hoepf, Christoph Schaeffer, Wolfgang Schäfer, Siegfried Stender

Fraunhofer-Institut für Produktionstechnologie, Germany (FhG IPT)
Head of the Institute: Fritz Klocke
Kristian Arntz, Jörg Frank, Christoph Neemann, Sebastian Nollau

Fraunhofer-Institut für Umwelt-, Sicherheits- und Energietechnik,
Germany (FhG UMSICHT)
Head of the Institute: Eckhard Weidner
Sylke Palitzsch, Hartmut Pflaum, Manuela Rettweiler, Uwe Schnell

Fraunhofer-Institut für Werkzeugmaschinen und Umformtechnik,
Germany (FhG IWU)
Head of the Institute: Reimund Neugebauer
Lars Georgi, Hans-Joachim Koriath, Ralf Lang, Frank Treppe

Institut für Industrielle Fertigung und Fabrikbetrieb, Universität Stuttgart, Germany
(Universität Stuttgart IFF)
Head of the Institute: Engelbert Westkämper
Carmen Constantinescu

Instituto de Engenharia de Sistemas e Computadores do Porto, Portugal
(INESC-Porto)
Head of the Institute: José Mendonca
Antonio Correia, Alves, Américo Azevedo, José Carlos Caldeira, Joao Jose Ferreira,
José Soeiro Ferreira, Ricardo Madureira, Antònio Lucas Soares, Jorge Pinho Sousa

Loughborough University – Wolfson School of Mechanical and Manufacturing
Engineering, United Kingdom (LBORO)
Head of Unit: David Williams
Paul Hourd, Kathryn Walsh

National Research Council – Institute of Industrial Technologies and Automation, Italy (CNR-ITIA)
Head of the Institute: Francesco Jovane
Andrea Ballarino, Carlo Brondi, Emanuele Carpanzano, Andrea Cataldo
Maria Stella Chiacchio, Daniele Dalmiglio, Cecilia Lalle, Giacomo Liotta
Augusta Maria Paci, Francesco Paolucci, Marco Sacco, Francesca Tiberi

TNO Knowledge for Business, Netherlands
Egbert-Jan Sol

Verband Deutscher Maschinen- und Anlagenbau e.V., Germany (VDMA)
Claudia Rainfurth, Dietmar Göricke

Wroclaw University of Technology, Institute of Production Engineering and Automation; Centre for Advanced Manufacturing, Poland (CAMT)
Head of the Institute: Edward Chlebus

Printing: Krips bv, Meppel, The Netherlands
Binding: Stürtz, Würzburg, Germany